ANIMAL BEHAVIOUR · VOLUME 2

COMMUNICATION

ANIMAL BEHAVIOUR

A SERIES EDITED BY

T.R. HALLIDAY
Department of Biology
The Open University

AND

P.J.B. SLATER
School of Biology
University of Sussex

ANIMAL BEHAVIOUR · VOLUME 2

COMMUNICATION

EDITED BY T.R. HALLIDAY

AND P.J.B. SLATER

BLACKWELL SCIENTIFIC PUBLICATIONS

OXFORD LONDON EDINBURGH

BOSTON MELBOURNE

© 1983 by
Blackwell Scientific Publications
Editorial offices:
Osney Mead, Oxford OX2 0EL
8 John Street, London WC1N 2ES
9 Forrest Road, Edinburgh EH1 2QH
52 Beacon Street, Boston,
    Massachusetts 02108, USA
99 Barry Street, Carlton,
    Victoria 3053, Australia

First published 1983

Photoset by
Enset Ltd, Radstock Road,
Midsomer Norton, Bath, Avon
Printed and bound in
Great Britain by
Butler & Tanner Ltd,
Frome and London

Distributed in the USA and Canada by
W.H. Freeman & Co., San Francisco

British Library Cataloguing in
Publication Data

Communication.—(Animal
behaviour; v. 2)
    1. Animal communication
    I. Halliday, T.R.   II. Slater, P.J.B.
    III. Series
    591.59   QL776

ISBN 0-632-00903-9
ISBN 0-632-00884-9 Pbk

CONTENTS

v

SERIES INTRODUCTION

As Niko Tinbergen, one of the founders of ethology, pointed out, if one asks why an animal behaves in a particular way, one could be seeking any one of four different kinds of answer. One could be asking about the evolutionary history of the behaviour: why did it evolve to be like it is? One could be asking about its current functions: through which of its consequences does natural selection act to keep it as it is? Thirdly, one might be interested in the stimuli and mechanisms that lead to the behaviour being performed: what causes it? Finally, one might be asking about development: how does the behaviour come to be as it is during the life of the individual animal? A complete understanding of behaviour involves investigation of all these questions, but in recent years there has been a tendency for ethologists to specialise in one or other of them. In particular, the functional analysis of behaviour has almost become a separate discipline, variously called behavioural ecology or sociobiology. This fragmentation of the subject is unfortunate, because all its facets are important and an integrated approach to them has much to offer.

Our approach in these books has been a more wide-ranging one than has been common in recent texts, with attention to all the kinds of explanation that have traditionally been the concern of ethologists. Aimed at students, each volume will provide a comprehensive and up-to-date review of a specific area of the subject in which there have been important and exciting recent developments. It is no longer easy for a single author to cover the whole field of animal behaviour with full justice to all its aspects. By asking specialists to write the chapters, we have tried to overcome this problem and ensure that recent developments in each area are fully and authoritatively covered. As editors, we have endeavoured to make sure that there is continuity between the chapters and that no significant gaps have been left in the coverage of the theme specific to each book. We hope that students who are inspired to further study by what they read will find the Selected Reading recommended at the end of each chapter a useful guide, as well as the more specific references which are gathered together at the end of each book.

We thank Bob Campbell and Simon Rallison of Blackwell Scientific Publications for their help and encouragement throughout the preparation of these books, Clare Little of Oxford Illustrators for her fine work on the illustrations and, most important of all, our authors for their readiness to accept a well-defined brief, to meet deadlines, and to accept our editorial changes and promptings.

T.R.H.

1983   P.J.B.S.

ACKNOWLEDGMENTS

Both of the editors and, in most cases, some of the authors of other chapters have commented on each chapter in draft. In addition, the authors of individual chapters would like to express their thanks to the following for their comments: John Krebs (Chapter 2); Murray J. Littlejohn and Geraldine C. Richards (Chapter 3); T. Szabo, A. Bass and D. Kemenovic (Chapter 4).

INTRODUCTION

The study of communication between animals has always been central to the interests of ethology. As with so much else in biology, animal behaviour received major impetus from the writings of Darwin, and his book *The Expression of the Emotions in Man and Animals* (1872) was mainly about communication. Another early study, which is also full of interesting and stimulating ideas, is that by Julian Huxley (1914) on the courtship of the great crested grebe. This paper is remarkable for showing what can be achieved with 'a good glass, a notebook, some patience, and a spare fortnight in the spring' as Huxley puts it, but also because many of the ideas of later ethologists can be seen germinating within it. More formal theories of ethology were also stimulated to a great extent by such work on displays as that of Konrad Lorenz (1941) on ducks and of Niko Tinbergen (1952a) on sticklebacks and, since then, research on communication has continued to be a major source of ideas about animal behaviour as a whole. The coming of age of ethology as a science can be said to have occurred in 1972 when Lorenz and Tinbergen shared a Nobel prize with Karl von Frisch, who had elucidated the dance language of bees. It was a prize as much for work on animal communication as for any other single area of the field.

One reason why animal communication has caught the interest of ethologists is that many of the signals produced by animals are as striking to human observers as they are to the other animals at which they are directed. The distinctive smell of a skunk or of a tomcat, the remarkable colour patterns of birds, butterflies and tropical fish, the facial expressions of monkeys and of people, the monotonous sounds of insects and the fantastically varied songs of some birds: all of these catch the imagination and demand explanation. In the tough world that animals inhabit, to do other than blend with the background is a dangerous enterprise, but the demands of communication and of camouflage are incompatible: just how has natural selection solved this dilemma and what is it about each act of communication that is so important that it has led

1

the animal to abandon its crypsis? The search for answers to such questions has given rise to a great deal of work, and to many of the delightful stories that make natural history on television such compulsive viewing, but this research has also had important scientific points to make, and these will be our primary concern in this book. As we shall see, there has hardly been an important idea or theory within ethology which has not been illuminated by studies of communication.

Why is this so? One reason is undoubtedly that the communication signals of animals are usually highly stereotyped. In many cases one of the most important pieces of information that they transmit is the species of the signaller. This is crucial, for example, when an animal is advertising for a mate. Such signals have to conform to precise rules common throughout the species: any individual that diverges from these rules will be best rejected by members of the opposite sex as possibly belonging to another species with which hybridisation would be disadvantageous. There are thus good reasons for stereotypy in signals, which do not apply to behaviour patterns such as feeding movements or those involved in grooming. This means that the displays of animals are comparatively easy to study, at least at a superficial level: they can be described precisely and recognised when they are repeated, and the context of their occurrence can be determined without too much difficulty. In the terminology of W. John Smith (1968), we can thus begin to determine what message they convey.

This makes the whole enterprise appear simple, but the difficulty is to get beyond these initial stages. Two main problems crop up here. The first is that animal signals usually convey only a very broad message about the motivation of the individual producing them, rather than the sort of precise points that our language allows us to make to each other with words. So, even if we do know the context in which a signal appears, it may be difficult to decide exactly what its message is and how this differs from the message of other signals. The fact that the same signal may occur in several different contexts and need not necessarily mean the same in all of them adds to the confusion! A second difficulty with studying communication is that it involves more than one animal. We have to look and see what effect signals have on other individuals as well as determining what they say about the animal that produced them. If we study feeding or grooming we can do so in a relatively quiet and controlled environment, the outside world kept as constant as possible so that our results can be easily interpreted. We simply cannot do this with communication because signalling depends on the presence of others, with all the hurly-burly that that involves.

Studying communication is not, therefore, as straightforward as it might seem or one might hope. This is a point that Slater brings out in Chapter 1, although he also shows that there is now a strong armoury of techniques which can be used to tease apart the various influences that are involved. In the best ethological traditions, some of these rely on observation alone, followed by various statistical methods to see whether the behaviour of different animals is correlated. Like Huxley, some ethologists still use only a notebook, but for others closed-circuit television and computer keyboards have become standard tools which allow behaviour to be analysed in far greater detail. Nevertheless, one can only get so far by observation: to decide whether a signal really does lead to a particular response, an experiment is virtually essential, and Slater also outlines the various possibilities here. While it may seem obvious to the casual observer that one animal is communicating with another, it is remarkable how difficult it is to demonstrate that the behaviour of one individual actually influences that of the other!

It is questionable whether demonstrating an effect of the behaviour of one animal on that of another is all that is needed to show that communication is taking place. In Chapter 1, Slater argues that it is not, though some of our other authors adopt rather different views of what communication involves. We have not attempted to push them into uniformity, for each of the various views deserves consideration and has arguments in its favour. Deciding what communication is or is not can come down to fairly fine distinctions, a point with which there is an interesting parallel in deciding whether or not a type of communication constitutes language. At one time it seemed clear what language was and how it differed from anything that animals could achieve, but then chimpanzees were trained to converse with humans using sign languages of various sorts. The most famous of these animals, a female called Washoe, mastered over 300 symbols in American Sign Language, and would string them together into rudimentary sentences. As the information on Washoe's capabilities accumulated, some linguists continued to object that she was still incapable of real language by citing ever more minor points which were true of humans but not of her. As one cynic observed, it would be simpler just to define language as 'what a chimp cannot do' and be done with it! With communication, as with language, it is easy to get sidetracked on to the semantic issues which are involved in defining words. This would be a shame for, whether or not Washoe has mastered a language or a particular behaviour pattern can be regarded as communication, the feats involved, and the problems they present, are real enough.

When animals communicate with each other, information of various different sorts passes between them. The word information is rather a confusing one because it has a specialist scientific meaning as well as its everyday sense; both of these are relevant here, as Halliday discusses in Chapter 2. In the general sense of the word, a signal may incorporate details of the species, age or sex of the signaller, the group to which it belongs, its family or even exactly which individual it is. The signal may transmit information about the outside world, as in the bee dance which indicates the location of food, or about the signaller's state of motivation, as when a threat display shows that it is likely to attack. These and many other points of information may be gleaned from the signals an animal produces. In the strict scientific sense, as defined in information theory, so many 'bits' of information pass from one animal to another as a result of a signal. The more bits, the more the uncertainty of the recipient as to what will happen next is reduced. Halliday also discusses this specialist usage, and it is considered further by Wiley in Chapter 5. Because information theory allows us to measure the amount of information that passes from one animal to another, it can also help us to decide whether this amount is greater than zero, in other words whether or not any communication is likely to be occurring at all, a use mentioned in Chapter 1.

Much of the communication of animals involves vision, olfaction or audition, although some species rely on other senses, such as the electroreception used by various fish. Which sensory modality is used for a particular signal depends to some extent on the information it conveys. In advertising for a mate, for example, larger animals tend to use sounds, as these travel rapidly, move round obstacles and can be detected at great distances. Smaller animals cannot achieve such high sound intensities and may rely instead on pheromones, smells which travel downwind and can, under favourable conditions, attract mates from remarkable distances. Obviously, odours cannot be used to signal rapidly changing states, as can gestures or sounds, but this may be an advantage in some cases. An animal which marks the boundary of its territory with a scent gland or with urine relies on the persistence of the smell to continue advertising ownership even when it is some distance away.

These rather general points concern the uses of the three main senses, but within each there are also important constraints on what signals are like, how their form is related to the function that they serve. This has been an especially active field of study in the past few years, and it is one of the major points taken up by Gerhardt in Chapter 3. The environment of any animal is a complex place and the problem is to produce a signal which will travel through it and reach recipients as efficiently and with as little distortion as possible. Studies of just how environments do differ in their transmission characteristics, in time as well as in space, have helped us to understand why some animal signals have the form that they do: the adaptations are often very precise and elegant. As Gerhardt explains, signals must take account of the animate aspects of the environment as well as its purely physical characteristics; they must diverge from those of other species if they are to stand out and be detected; they should also be as hard as possible for predators to spot and home in upon. With such forces pulling in different directions, the form of signals must usually be a compromise rather than a perfect solution to any one problem. Yet the neat way in which some signals are adapted to their function is extraordinarily impressive: the female fireflies which imitate the sex attractant signal of their prey species and make a meal of the males that approach is just one of several examples discussed by Gerhardt.

Having dealt with the messages encoded in signals and the form of the signals themselves, we move in Chapter 4 to an account of the sensory aspects of communication. Here Hopkins discusses how the sense organs and brain deal with incoming signals and extract from them information on such questions as whether or not there really is a signal present amongst all the other stimuli being received and, if there is, which out of all the possible signals it is and where it is coming from. In many species, sense organs are used for a variety of purposes, such as the detection of food and of predators, as well as for receiving signals and passing them on to the brain. Appropriate to this, these animals have sense organs which respond to a wide range of stimuli and thus are not especially well matched to any one of them. The recognition of one particular stimulus cannot just be because the sense organs are tuned to this alone. However, in other cases which Hopkins describes, the sense organs are specifically adapted to receive particular stimuli. The tuning of the bullfrog ear to bullfrog calls is one example and another, though perhaps not strictly speaking lying in the realm of communication, is the greatly enlarged eye of many male flies which enables them to detect and pursue females.

The final chapter takes a rather more functional approach to communication. With the growth of interest in how natural selection has led to particular adaptations, the evolution of communication has become an especially exciting and fruitful area of study, as Wiley makes clear. The now widespread acceptance that selection acts on the individual, rather than on the group or the species as has often been supposed, has important implications here. It suggests that one animal will only transmit accurate information to another if it is in its own best interests to do so. Whether the recipient benefits is irrelevant, except to the extent that genes are shared between the two animals. In line with this, Dawkins and Krebs (1978) have argued that it is better to view transmission of a signal as one animal manipulating another to its own advantage, rather than as a sharing of information. This claim sounds somewhat sensational, but whether it is more than playing with words remains to be seen.

Another advance in evolutionary biology that is especially relevant to communication is also a theoretical one: this is the development of the concept of an 'evolutionarily stable strategy' or, to be less cumbersome, an ESS, an idea originally put forward by John Maynard Smith in 1972. He considered fights between animals and wondered why, if they were concerned only with their own interests, the individuals often simply displayed to each other rather than coming to grips and tearing one another to shreds. His theoretical models showed that displaying and only fighting if one's opponent did so could actually be an ESS, a strategy which could not in the long term be bettered by any other.

Many different models based on this idea have followed and, while they tend to be too simple for many natural situations, they have suggested reasons for hitherto unexplained results in some areas and made useful predictions in others. While largely theoretical at present, this is an active and exciting area of research, much of it concerned with communication: as a result there is hardly a paragraph of Wiley's chapter that could have been written ten years ago.

This book, therefore, looks at communication from several different viewpoints: the basic methodology in the first chapter, the production, transmission through the environment, and reception and analysis of signals in the next three, and finally the evolution of communication in the fifth. We hope to have shown that communication has relevance to a wide range of studies in animal behaviour. Even the bright colours of flowers which attract bees to pollinate them, or the secretions of protozoa which influence the movement of others of their kind, are examples of organisms communicating with one another. But if communication is relevant in one field of study more than in any other it is in the area of social organisation and behaviour, for social relationships are built up from interactions between individuals. Animals vary enormously in their social structures, even where the species involved are very closely related. Such contrasts may stem from tiny differences between species in ecology and behaviour; for example, Kummer (1971) argues that the sharp distinction between the one-male group of hamadryas baboons (*Papio hamadryas*) and the larger groups found in the anubis baboon (*P. anubis*), in which there are no long-term bonds between individual males and females, stems from the simple fact that males herd females in the former species but not the latter.

Even if social structures do vary between species it is easy to be impressed by how organised animal groups tend to be, with dominance hierarchies, for example, being commonly found. Given the evolutionary argument that each individual should act for the benefit of its own genes, how can such organisation emerge? If animals live together for a long time and can recognise each other as individuals, they can clearly cooperate and develop relationships which may vary from pair to pair depending on their past experience of communication with each other. In many species individuals of one of the two sexes stay together in family groups while those of the other disperse: most mammalian groups are, for example, matriarchal, with females staying within them and males moving elsewhere. This too may affect social relationships within the group, with the reactions of individuals towards one another being influenced by the extent to which they share genes and signals having evolved to convey the relevant information. These arguments show how communication in social groups can be a complicated affair, but they do not get beyond the fact that individual animals in social situations are still acting as individuals: any structure or apparent organisation in the group is not in itself a product of natural selection, but is an emergent property of the relationships between individuals, each selected to maximise its own inclusive fitness. Perhaps the most challenging task for studies of animal communication in the future is to understand how these structures do emerge from series of interactions between pairs of individuals such as the acts of communication with which this book has been largely concerned. The field of animal communication is a very active one and, whether one seeks fascinating stories or deep theoretical insights, it is a cornucopia which shows no signs of becoming exhausted.

CHAPTER 1
THE STUDY OF
COMMUNICATION

P.J.B. SLATER

1.1   What is animal communication?

In a book on communication it is quite possible to devote a great deal of space to discussing the meaning of words. This would not be helpful. However, it is worth paying a little attention to the word 'communication' itself, for this has a number of different meanings, not all of which are appropriate to the study of communication between animals.

From its derivation, communication can be taken to mean the sharing of anything between A and B, as, for example, in the expression 'communicating rooms'. Communications engineers use the term in a slightly more restricted sense to cover 'the transmission of information regardless of its origin or destination' (Mackay 1972). Communication, in either of these senses, may occur between two inanimate objects, or between such an object and an animal which perceives it. Clearly, these uses are too general for our purposes: the essence of animal communication is that one animal influences another in some way. Even here, however, there is room for disagreement. Marler (1967) rejects the notion that a mouse rustling in the grass is communicating with the owl that hunts for it, on the grounds that the mouse has certainly been selected for avoidance of the production of such stimuli. He suggests that the 'synergistic interplay between participants, both of which are committed to maximising the efficiency of the interchange' is an essential feature of animal communication. A great many examples do indeed fall into this category, but there are some which do not, and the most striking of these are some of the cases where communication takes place between species. For instance, where a mother bird shows a distraction display, behaving as if she had a broken wing and so

9

leading a predator away from her chicks, she gains but the predator loses. The same is true in such cases of mimicry as that in which a tasty butterfly mimics an unpalatable one and so avoids being eaten. Here again the signal serves to mislead the recipient.

There are undoubtedly also cases where animals mislead one another within a species so that the signaller benefits at the expense of the recipient of the signal. Dawkins and Krebs (1978) have gone so far as to suggest that communication is more a matter of one animal manipulating another than of information transmission. While it is worth stressing that manipulation or deceit may be involved, as this point has often been neglected (for example by Smith (1977)), their emphasis on this point is overdone (Hinde 1981). For an example of communication to have evolved it is essential that signal transmission should, on average, be advantageous to the sender. In some cases this advantage may arise because the recipient is misled, but in the great majority the recipient will benefit also, the sharing of information helping to integrate activities between mates, social companions and territorial neighbours and between parents and their offspring. In close-knit groups of animals, where individuals recognise one another and interact with each other over extended periods, the long-term penalties which may arise if deceit is discovered may more than outweigh any short-term gain it makes possible. Nevertheless, if selection is to lead an animal to produce a signal, some benefit must accrue to that individual and, if it does, the signal will still be produced whether or not recipients gain also. Thus animal communication can be defined as 'the transmission of a signal from one animal to another such that the sender benefits, on average, from the response of the recipient'. By avoiding use of the word 'information' we escape from the implication that communication must always be truthful: a signal can be misleading as well as helpful.

Rather than suggesting that the sender must gain through natural selection if a signal is to be transmitted, earlier definitions have often made a similar point by stressing the role of intention in animal communication. Unlike the mouse in the grass, the sender must intend to transmit the signal before communication is said to occur. While this may be a useful criterion for humans, who can be asked about their intentions, it is not easy to establish criteria for intention in other animals (Green & Marler 1979). A definition expressed in terms of the advantage of communicating rather than its causes therefore seems preferable.

One problem with a definition depending on advantage to the sender is that this advantage must often be assumed to exist rather than empirically demonstrated. In many instances it is not easy to show that the sender gains from communicating and some signals, such as alarm calls, may even appear at first sight to be disadvantageous. But wherever an animal produces a sight, a sound or a scent which causes it to stand out from the background rather than blend with it, there must be strong grounds for supposing that communication is involved, even if the advantage is not obvious. If the supposed signal is stereotyped and has a distinctive structure, and particularly if the animal has an organ such as a scent gland which is specially adapted for its production, then the evidence is so much the better. If a signal is suspected then just what the advantage is to the signaller can be a subject for study: in those cases which have been examined it has usually been possible to suggest one which fits the observations even if firm evidence is hard to obtain. Alarm signals are a good case in point; while they appear to be altruistic, there has been no shortage of suggestions as to why their production is beneficial to the signaller (Harvey & Greenwood 1978). Another tricky case involves signals which apparently attract other individuals to food (Wrangham 1977). In close-knit social groups, where members are likely to be related to each other, this may be beneficial. But it has also been suggested (e.g. by Nelson (1979)) that some seabirds have white plumage because this attracts others to the fish shoals on which they are feeding. It is hard to see an advantage to the individual in being attractive in this way although, if there is enough food for all, the disadvantage may be negligible. It seems more likely, as others have argued (e.g. Craik 1944), that the white plumage enables the predator to get closer to its underwater prey, so that neither altruism nor a signal is involved.

A final difficulty concerns those cases where the behaviour of one animal influences that of another without a signal as such being involved. Thus animals in groups often do similar things at the same time, and it can be shown experimentally that this sometimes arises through social facilitation, one individual starting to bathe or feed leading others to do the same shortly thereafter (e.g. Birke 1974). Various different mechanisms may be responsible

(Tolman 1968; Clayton 1978), but in all cases the behaviour of one individual is influencing that of others. Sometimes a signal, such as a special call produced when feeding (Clifton 1979), may be involved, but often the synchronising activity is clearly adaptive in other ways without having any attributes to suggest that it is adapted to provide a signal. Nevertheless, synchrony is probably advantageous to all members of the group. Predators are unlikely to spot a group of animals all of which are sleeping in cover. If all move and feed at the same time, none will get left behind and each will be able to avoid predators through both its own vigilance and that of others. Hence the animal whose lead is followed benefits as well as the rest. Social facilitation therefore does involve the behaviour of one animal influencing that of others, and it is also likely that the first animal benefits from the synchrony which results. But, by the definition given above, it does not qualify as communication unless a special signal is involved. Many people might disagree with this and it certainly highlights the problem of definition. Here one has to differentiate between an animal which bites at food in the normal way and one which does so in a slightly exaggerated manner, thus producing a signal which enhances social facilitation.

All these examples illustrate the difficulty of arriving at an entirely satisfactory definition of communication which is unambiguous and can easily be applied to all cases. But they also give an indication of the breadth of the subject and the variety of ways in which one animal can affect another's behaviour. Whatever definition is used, some will disagree with it. But, regardless of semantics, for the great majority of cases where the behaviour of one animal influences that of another, all will be agreed upon whether or not communication is involved. The rest of this chapter is concerned with ways in which communicative acts can be analysed. The next section deals with the component parts into which they may be split, and the final two sections consider how observation and experiment respectively can help us to understand what is going on when animals communicate with one another. The field is a large one; we shall review the many different ways in which people have studied communication with a few examples to give some indication of what these approaches have shown.

1.2   Analysing a communicative act

1.2.1   *Messages, meanings and contexts*

An act of communication can be viewed from two different vantage points: that of the sender and that of a recipient. This point has been made most clearly by Smith (1965, 1968, 1977), who makes the important distinction between *message* and *meaning*. The message is what the signal encodes about the sender and in some way describes the state of that individual. It may, for example, indicate that the individual is anxious, that it is an adult male in breeding condition or that it has just seen a predator. The meaning, on the other hand, is what the recipient makes of the message. This can only be inferred from the response, if any, that the recipient shows. It can differ very greatly from one recipient to another and is quite distinct from the message. The meaning can also vary according to context so that the same signal may elicit different responses in different circumstances. Because of this a comparatively small repertoire of signals may come to elicit many different responses, the one shown depending on the nature of the individual perceiving the signal and on its experience, as well as on the immediate context and on variations in the signal itself. Subtlety of communication does not necessarily demand a huge number of different signals.

The distinction between message and meaning can best be explained by reference to one or two examples. Consider a male bird singing at the top of a tree (Fig. 1.1). In many bird species, only males sing, they only do so during the breeding season, they require testosterone to be circulating in their blood, and they sing much more when they are unmated (Catchpole 1979). The message in a particular species' song may therefore be 'I am an unmated male *Dissimulatrix spuria* in breeding condition'. The meaning of the song will vary according to the listener. To a predator it may mean 'Approach stealthily and attempt to catch'; to a songbird of another species it may simply mean 'Ignore'; to a male of the same species it may mean 'Keep out. This territory is already occupied'; to an umated female it may mean 'Approach and attempt pair formation'. The identity of the listener and the context determine the response that will be shown, and from this

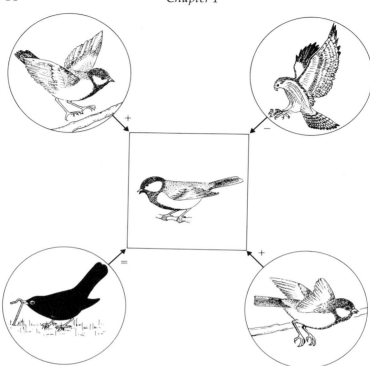

Fig. 1.1. The advantages and disadvantages of communicating. The male bird in the centre is singing on his territory. Advantages accrue to him if his song attracts a female (top left) or repels rival males (bottom right). His song may be ignored by other species (bottom left) but may also attract predators (top right), which is a possible disadvantage of singing. The production of a signal will only be selected for if the net advantages outweigh the disadvantages.

the different meanings that the signal can have may be inferred. Some of these meanings may actually be disadvantageous to the sender, as when a predator is attracted. But, for the signal to be produced, the net result must be advantageous, as was stressed in the previous section.

When an animal communicates, the message is encoded in a signal. Signals are often rather stereotyped, although they may form graded series so that slight nuances can be transmitted. Whichever is the case, the message in some way defines the probability of what the signaller is doing or may do. Put differently, it results from the signaller being in a particular state. It is often not easy to define this state as the signal may appear in many different contexts and animal signals are often more closely linked to states of motivation or emotion than to situations which are easily put into words. The following examples will indicate some extremes.

Vervet monkeys (*Cercopithecus aethiops*) have several different alarm calls which may be produced when predators are spotted (Struhsaker 1967). Three of these carry very precise messages linked to the nature of the predator. Their messages are quite simple: 'I have seen a snake', 'I have seen a leopard', 'I have seen an eagle'. Their meanings are also appropriate and unambiguous. Other monkeys respond to the snake call by looking downwards or approaching and examining the snake. To the eagle call they respond by looking up and seeking cover in the undergrowth, while the leopard call leads them to climb hurriedly into trees (Seyfarth *et al.* 1980). The system is so precise that the signaller might almost be shouting the name of the particular predator.

Such a clear correspondence between messages and their meanings is unusual. A much more complex and very interesting case is that of a call produced by the eastern kingbird (*Tyrannus tyrannus*), a flycatcher widespread in North America (Fig. 1.2). This call, the 'kit-ter', occurs in a number of different circumstances described by Smith (1968). A male will produce it when he

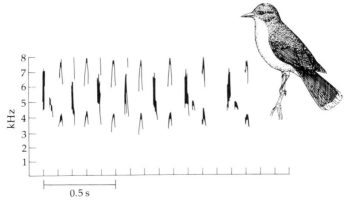

Fig. 1.2. The male eastern kingbird and a sonagram of the 'kit-ter' vocalisation produced by this species (after Smith 1977). Sonagrams are plots of sound as frequency in kHz (1 kHz = 1000 cycles s⁻¹) against time.

stops flying during a patrol of his territory or as he turns having chased off a hawk. Birds of either sex may call in this way, particularly early in the season, when they approach their mate, and well-grown young may give the call when approaching their parents to beg from them. In both these situations the animal is moving into the close proximity of another individual which could possibly attack it. Finally, a female may give this call if she leaves her nest when her mate is absent: normally he returns and guards it while she is away. At first sight all these situations are quite different from each other, but Smith (1968) argues that they have in common the fact that the individual is uncertain whether or not to move. He labels the call the Locomotory Hesitance Vocalisation or LHV and considers uncertainty about moving to be its message. What then of its meanings? Rather like song it may act to exclude males from and attract females to an area. But it may also have influences within a mated pair. For example, a newly paired male hearing his mate call as she approaches him may take the meaning as 'Don't attack or I will fly away'. Later in the season the same male hearing the call may respond quite differently, taking the message as 'Go to the nest and guard it'. This example stresses the importance of context: the same animal in two different situations may react quite differently to a signal. The message is always the same but the meaning varies.

1.2.2   Signals

The message is what a signal says about its sender, the meaning is what the receiver extracts from it. Neither of these is tangible; they cannot be studied directly but must be inferred from behaviour. On the other hand, the signal is the physical form in which the message is embodied for its transmission through the environment.

For most animal signals, the form of the signal has no clear relationship to the message that it encodes, just as the word 'dog' does not look or sound anything like the animal to which it refers. However, signals are not absolutely arbitrary. The modality in which they are transmitted (Chapter 4) and the form of the signal within that modality (Chapter 3) may both be determined by considerations of how the signal is most effectively transmitted given its function and the need to avoid the attraction of predators.

A well-known example of this is the 'seeep' alarm call of passerine birds, found in very similar form in many species (Marler 1955) and which even has equivalents in some mammals (see, for example, Owings & Virginia 1978). Thin, high-pitched whistles like these have probably evolved because the sources of signals of this form are especially hard to locate and this is important in a call produced when a predator is present. Conspecifics do not need to know where the signaller is in order to react appropriately, and it is essential to the caller that the chances of the predator locating it are minimised. Hence the calls of many species have converged on the same signal form, and these species will react to each other's signals, without it necessarily being advantageous for an animal of one species to tell those of another about the danger. There are other cases in which the form of a signal seems to be related to the message that it conveys. Morton (1977) argues that this is so with many bird and mammal sounds, harsh, low-pitched calls being used in hostile contexts and purer, higher pitched ones where the situation is friendly. He suggests that the former may be intimidating because harsh sounds of low frequency can only be produced by large individuals. Thus animals might be expected to avoid those producing deeper sounds than themselves, as size is such an important factor in success in fights. Selection should then lead to aggressive signals being as harsh and low as possible. It may favour a contrast to this in appeasing and friendly sounds to make them easy to discriminate and so to minimise the chances of their eliciting a hostile response. That signals of opposite meaning are also often opposite in form, presumably to avoid ambiguity, was first noticed by Darwin (1872), who referred to it as the principle of antithesis; Morton's idea is similar. Darwin himself pointed to the extreme difference in posture between a hostile dog and the same animal 'in a humble and affectionate frame of mind' (Fig. 1.3). A similar contrast is seen in gulls, where the beak is shown off in aggressive displays but is hidden, by turning the head away, in appeasing ones (Tinbergen 1959).

These examples show that signals used to convey similar messages may have features in common. Often this will be because the best possible form of signal for a particular function is moulded by environmental constraints but, in some cases, as Morton suggests, the form of the signal may be related to the content of the message itself.

Fig. 1.3. An example of the principle of antithesis used by Darwin. The same dog is portrayed above 'approaching another dog with hostile intentions' and below 'in a humble and affectionate frame of mind'. (From Darwin 1872.)

1.3   Observing communication

Most of the best ethological studies start with a period of observation during which the observer becomes familiar with the animals and formulates hypotheses for later testing. Many studies consist entirely of observation for, as we shall see, it is often difficult to devise experiments to examine communication without disrupting the very behaviour one is trying to study. Just how far can the study of communication be taken by observation alone? Once again this question can be tackled from two different viewpoints: that of the sender and that of the receiver. We can look at the sender's behaviour and ask what causes it to produce a signal (in other words what message a particular signal encodes) or we can see how a recipient responds and try to discover from this the meaning that the signal has for it.

1.3.1 What is the message?

Although they are not absolutely fixed in form, animal signals are often discrete, all-or-nothing events which are more stereotyped than non-signal movements (Barlow 1977). They are thus relatively easy to classify and identify. However, some signals do vary in intensity. An oft-quoted example of this is crest raising in the Steller's jay (*Cyanocitta stelleri*), the height of the crest in an aggressive encounter being greater the more fiercely the animal is being opposed (see Fig. 1.4; Brown 1964). It is also true that signals may show a spectrum of subtle variations, so that a continuum of

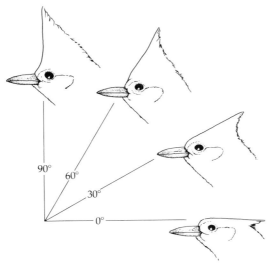

Fig. 1.4. A graded signal: crest elevation in Steller's jay. The angle of the crest correlates with the behaviour shown by the bird. (From Brown 1964.)

intermediates may be found between extremes that are quite distinct. The sounds of rhesus monkeys (*Macaca mulatta*) are a good example of this, as many of them grade into one another (Rowell & Hinde 1962). But, notwithstanding exceptions of this sort, many animal signals can be regarded, at least at the outset of a study, as if they were invariant. Analysis of the variations in signals, and relating these to differences in the messages they encode, may be crucial to a full understanding, as for instance where the variation occurs largely between individuals and so may encode their identity (e.g. Stamps & Barlow 1973; Marler & Mundinger 1975), but more general functions of the signal can be found out without it. There are theoretical reasons why many signals vary rather little: their fixity may lead them to be unambiguous (the notion of 'typical intensity' put forward by Morris (1957)) or it may enable individuals to disguise their motivation (Maynard Smith 1972; see Chapter 5). But whatever the reason for stereotypy, its main interest here is that stereotyped signals are much easier to study!

Having defined the signal in question, the way to discover the message that it conveys is to study the contexts in which it occurs. Many ethologists have done this rather informally, as with Smith's study of the 'kit-ter' vocalisation mentioned earlier, the sorts of situations in which the behaviour occurs being described in an attempt to determine what they have in common. Indeed, in the case of Smith's study, more formal analysis might have been fruitless because the behaviour occurs at such a variety of times and places. Much of the classic work of Tinbergen (1959) on gulls was also done by informal means and, despite this, it certainly shed a great deal of light on the messages of their displays. But there is a limit to the extent that others can take on trust the 'judgement of the skilled observer' when it is on this basis alone that the role of a behaviour pattern has been assessed. Ultimately, precise quantitative analysis is called for, both to achieve a detailed understanding and to present results which may be appraised and repeated by others.

The commonest method used in the formal analysis of messages is the study of sequences. It is an assumption made here that 'a temporal correlation between movements indicates that they are somehow causally linked' (Hinde 1970, p. 199). Thus, by examining the sequences in which different actions are performed, the relationship of a signal to the rest of the animal's behavioural repertoire may be discovered. A particular signal may often be followed by attack, for instance, and this may be taken to indicate that its message is an aggressive one. However, signals are not often associated unambiguously with one overt action in this way: for example they frequently indicate an equal likelihood of incompatible actions such as attack and escape, a fact that led to the conflict theory of motivation (see Baerends 1975; Chapter 2).

To say that two behaviour patterns occur close together is not very illuminating unless one can show that they do so more than one would expect by chance. If an animal spends 90% of its time fighting, it is only useful to think of signals as being associated with this behaviour if they occur in the context of fighting significantly more than nine out of ten times. The detection of associations thus requires comparison with some sort of random model if it is to say anything useful about the contexts in which signals occur. Table 1.1 shows the commonest way of doing this, using hypothetical data from an animal which, for the sake of space, is assumed to have only five behaviour patterns. The number of times each type

Table 1.1. A complete matrix of transitions between five different behaviour patterns. The matrix is a simulated one and the lower figures in each cell are expected values based on the hypothesis that all acts follow each other at random. An overall x^2 test is significant ($x^2 = 322.8$, d.f. $= 16$, $p < 0.001$), allowing rejection of this null hypothesis. Asterisks indicate where main departures from randomness lie, assessed using a 2×2 x^2 for each possible transition (the four cells being A→B, A→Not B, Not A→B, Not A→Not B); $p < 0.05^*$, $p < 0.01^{**}$, $p < 0.001^{***}$.

	Signal A	Signal B	Bite	Flee	Groom	Row totals
Signal A	61	22	26	48	16	173
	41.9***	29.0	23.5	40.6	38.0***	173.0
Signal B	14	22	30	70	10	146
	35.3***	24.5	19.8**	34.3***	32.1***	146.0
Bite	6	24	25	44	2	101
	24.4***	17.0*	13.7***	23.7***	22.2***	101.0
Flee	47	33	0	0	46	126
	30.5***	21.1***	17.1***	29.6***	27.7***	126.0
Groom	45	19	16	6	83	169
	40.9	28.5*	22.9*	39.7***	37.1***	169.1
Column totals	173	120	97	168	157	715
	173.0	120.1	97.0	167.9	157.1	

of action is followed by each other one is entered in a transition matrix. To see whether sequences are random, the expected number of transitions from A to B is calculated on the assumption that all transitions are equally likely. This is easily done: the expected values for each cell in the table are worked out as

$$\frac{\text{(row total for A)} \times \text{(column total for B)}}{\text{grand total}}.$$

Usually, as in Table 1.1, some transitions are found to be commoner than expected and some rarer. A chi-square test can be used to check whether the whole matrix departs significantly from random and, if it does, this or the binomial test has often been applied to different cells to see where the non-randomness lies (e.g. Stokes 1962; Weidmann & Darley 1971), although there are now better methods of examining the contribution of particular cells to chi-square (Fagen & Mankovich 1980).

Most studies of sequences have assumed that this form of analysis is a reasonable approach (e.g. Baylis 1976; Boer 1980). But there are hidden pitfalls, discussed in detail by Slater (1973). Just which behaviour patterns should one include in the matrix? Everything the animal does down to the merest twitch or eyeblink? Deciding what to include or not to include can make an enormous difference to the findings, because adding something or leaving it out will affect the grand total and so the expected number of each transition. Inclusion of different behavioural situations will also affect the outcome. For example, if encounters between rival males and those between male and female are put in the same matrix, transitions between courtship acts will all appear to be commoner than expected, as will those between aggressive acts. Balancing this will be the fact that courtship acts are seldom followed by aggressive ones and vice versa. But these results may occur for the trivial reason that one situation encouraged courtship and the other aggression, without indicating anything about the way behaviour is organised. The results of Wiepkema (1961), in which he tested male bitterlings (*Rhodeus amarus*) in three different situations and found three different clusters of acts, may well be just such a case. The moral is that it is a mistake to analyse behaviours drawn from varied situations together, unless the aim is simply to show that the situations are different. Another pitfall is the 'major diagonal' problem, that of transitions between behaviour patterns and themselves. How does one decide when A stops and then starts again? Is an act of walking just one step, in which case the walk–walk cell will have a large entry, or is it everything that occurs between walking commencing and being succeeded by another activity, in which case walking will never follow itself? Clearly, the distinction is an arbitrary one, but the findings will be greatly affected by how it is made. The best answer to this particular problem is not to consider that sort of transition at all but only those between different behaviour patterns (Slater 1973).

Table 1.2. The same matrix as in Table 1.1 but excluding transitions between behaviour patterns and themselves. The expected values have been calculated to sum correctly for rows and columns following Goodman (1968). As before, the overall χ^2 is significant ($\chi^2 = 153.4$, d.f. $= 11$, $p < 0.001$) and the significance of individual cells is indicated, as in Table 1.1, by asterisks.

	Signal A	Signal B	Bite	Flee	Groom	Row totals
Signal A	—	22	22	48	16	112
		26.6	17.4*	49.6	18.2	111.8
Signal B	14	—	30	70	10	124
	32.2***		18.8**	53.4***	19.6**	124.0
Bite	6	24	—	44	2	76
	18.2***	16.2*		30.4***	11.2**	76.0
Flee	47	33	0	—	46	126
	40.8	36.4	23.8***		25.0***	126.0
Groom	45	19	16	6	—	86
	20.8***	18.6	12.2	34.6***		86.2
Column totals	112	98	72	168	74	524
	112.0	97.8	72.2	168.0	74.0	

Table 1.2 shows the same data as in Table 1.1 but excluding these transitions. The expected values for the off-diagonal entries have been calculated so that they add up correctly for both rows and columns yet, like the observed figures, none appears on the diagonal. This calculation is not nearly as easy as that of the expected values in Table 1.1; it involves successive approximation and is best carried out by computer. These expected values are based upon the assumption that *bouts* of each type of act are

randomly distributed in relation to each other. Statistical testing can be carried out in the same way, but degrees of freedom are fewer (if r is the number of rows, there are $(r-1)^2$ degrees of freedom for a complete matrix, but $r^2 - 3r + 1$ for that with the missing diagonal; see Lemon & Chatfield 1971). The differences between these two matrices are most clearly shown when their results are summarised in flow diagrams, a technique often used to illustrate sequences of behaviour. The diagrams in Fig. 1.5 show all the transitions from Tables 1.1 and 1.2 that are significantly more frequent than expected. Some of the transitions in the two diagrams are the same so that relationships between acts appear similar. For example, Signal B is associated with Biting and Fleeing in both charts, a point which might suggest that it is a threat posture. But there are also some very striking differences which show just how much the method of analysis may affect the results. The relationships of Signal A are a case in point. The first analysis suggested that it tends to follow Fleeing and that it is produced in bouts. However, once the bouts are excluded from the analysis by removing the diagonal, this signal appears instead to follow Grooming and precede Biting, suggesting a totally different context. Although the technique used in Table 1.2 is the better one, even this may lead to some bias. Notice that Fleeing is never followed by Biting in Tables 1.1 and 1.2, and that this is a highly significant effect. This result may be very simply because an animal that has run away is, by definition, some distance from its partner

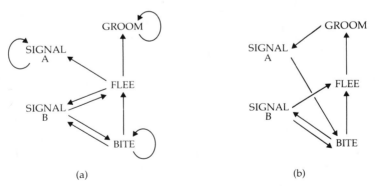

Fig. 1.5. Sequence diagrams from the data in Table 1.1 (left) and in Table 1.2 (right). Arrows are shown where the transitions indicated occurred significantly more often than random expectation.

and thus not in a position to bite it. If this is so it might therefore be more realistic to exclude the possibility of that transition from the analysis, just as the diagonal was excluded; this too can be done using the methods outlined by Slater (1973).

Finding out which behaviour patterns are associated with it may suggest the message of a signal, but being certain of this message is not an easy matter. A signal may precede one action most often, primarily because it inhibits another: showing a significant effect does not necessarily indicate that a signal was caused by the behaviour which preceded it. Furthermore, statistical significance is not a very useful criterion of whether one action is affected by another unless the relationships between the behaviour patterns being studied remain constant through time: in statistical jargon, the sequence must be stationary. This is often not the case in behaviour, the probability of one action following another changing with time; for example, it is likely to do so in a courtship sequence as copulation approaches. One good reason why it may change is that the acts of communication themselves do not just influence the next behaviour shown by the partner but have longer term effects on the partner's motivation (such influences are called 'tonic communication' by Schleidt (1973)). The following sequence is a non-stationary one: BABABABA-BCBCBCBCACACACACB. If the whole sequence were put in a matrix it would appear random, each of the three activities follows each of the others four times, but looking at shorter segments shows that this is far from being the case—it is just that the rules change during its course. Why not just analyse shorter sequences? This may be one answer, certainly, but another problem may lead to biases where samples are small. In the above sequence, A leads to B and B to A several times early on and, if they alternated enough times we might find that these two transitions occurred significantly more often than expected. Does this mean that they influence each other? Unfortunately it does not. Behaviour A might be courtship and B head scratching in a male bird which happens to have badly displaced feathers on his head but is trying to court his mate. Every time there is a break in courtship he scratches his head, so that the two behaviours alternate, but courtship does not cause scratching nor scratching courtship even though transitions between them are common. The only way to overcome this problem is to have a data sample large enough that influences of this kind even out: the coincidence that the bird happened to have an itchy head one day when courting is balanced by the fact that it was itchy again on another day, when he was fighting.

Thus large data sets tend to be biassed in one way, by non-stationarity, while small ones are biassed in another, by sampling errors. The best way to resolve this difficulty is to use a large sample, as long as the rules are the same throughout (one can test for this by splitting the sequence in two and analysing each half separately to see if they give similar results), or to combine a number of small samples from the same situation, for example looking at early courtship in several animals rather than the whole of courtship in one.

Sequence analysis is not the only way to assess the relationship between signals and the other behaviour patterns of an animal, but it is the most usual, and more detailed analyses are often based upon it. Other measures of association can involve the calculation of correlations over short periods (e.g. Andersson 1974). Moving on from measures of association between pairs of acts, the ways in which many different behaviour patterns may be interrelated can be examined using factor analysis (e.g. Wiepkema 1961) or cluster analysis (see Dawkins 1976a; Morgan *et al.* 1976). Information theory analysis, which is formally equivalent to chi-square (Chatfield & Lemon 1970), is another useful way to assess the influence of one behaviour pattern on another when both are performed by the same individual but also when the first is performed by one animal and the second by another, the topic of the next section.

1.3.2   *What is the meaning?*

Sequence analysis can help to establish associations between different behaviour patterns in the repertoire of an individual and thus indicate what the messages of the signals it shows may be. Similar techniques can be used in the study of meanings, but here the sequences examined are between animals rather than within an individual. After the first animal gives a particular signal what does the second animal do?

In some ways this is simpler. The results may be put into a matrix just like Table 1.1, the acts of one animal being listed down the side and the following ones of the other along the top. These need not be the same as each other (for example, the repertoires of male and female may be quite different if courtship is being studied) but, if they are, there is no reason to treat transitions lying on the diagonal as different from any others, so expected values can be worked out quite simply as in Table 1.1. Unfortunately, however, some other factors lead to difficulties in data collection, in analysis and in interpretation.

The major difficulty in data collection is that animals do not alternate behaviour politely, like people holding a conversation. For much of the time both may be performing actions simultaneously and each may be influencing the other. A hard and fast rule, appropriate to the speed of interaction, has to be chosen so that only transitions which comply are scored. One such rule might be: what is animal A doing at the moment when animal B starts to perform a particular action? Instead of this, if signals are discrete and responses clear, it may be better to record the first action of B after each of the possible signals given by A. When it comes to analysis, the main difficulty is one that can also influence studies of sequences within the behaviour of the signalling animal: the action that an interacting animal shows depends partly on signals received from its partner, but it also depends on what it has itself just done. The two influences are not easy to tease apart and the dependence of successive acts within each animal may lead to the appearance of an influence between them where none exists. If one animal happens to be grooming while the other is singing, for example, the two activities will appear to influence each other as a great many transitions will be scored between them even if the two animals cannot see or hear each other! As before, a large sample size is essential to ensure that this sort of effect is minimised. Finally come problems of interpretation. If it can be shown that one animal performing activity X tends to be followed by the other doing Y, does this show that X causes Y in any useful sense? That one causes the other is certainly a possibility but, going by observation alone, we must be cautious. All we have found is a correlation between the two actions, and arguing from correlation to causation is dangerous. Other interpretations are equally plausible. If X and Y also occur close together in the behaviour of one animal they may be acts with similar underlying causes; synchrony between animals, social facilitation and a similar environmental situation are all reasons why the two would occur close together in different individuals. These reasons would all lead X by one animal to follow Y by the other as often as Y followed X. Causation is much more likely if one of these transitions is much commoner than the other, but it is still not certain. Some influence, perhaps an earlier behaviour pattern by one of them, perhaps an environmental circumstance, might lead one animal to perform X with short latency and the other to perform Y with greater delay. Thus Y would often follow X without an influence of one on the other, like the chiming of two church clocks one of which is running slightly slower than the other (Cullen 1972).

The fact that most communicative systems are mutual, with two or more animals involved and signals and responses following each other thick and fast, is one reason why analysis is difficult. If a communicative encounter is prolonged, what each animal does may depend on changes in motivation during its course. But even in the short term, the two different influences we have discussed in this section and the last may combine to affect what it does. These effects are not easy to separate (Bercken & Cools 1980) and are often confounded in behavioural research by putting within- and between-animal sequences in the same matrix (Altmann 1965; Baylis 1976). As quite different mechanisms are likely to be involved in the flow of behaviour shown by one animal and the pattern of communication between animals, this is not very helpful. The most useful way to tease the two effects apart is using information theory (see Dingle 1972; Losey 1978). This can give a measure of the extent to which the behaviour of an animal may be predicted from its own past actions and of the extent to which it is influenced by signals received from others (see section 2.11).

It is remarkable how much insight into behaviour may be gained by observation alone, and many of the messages and meanings of signals may be rightly suspected when one watches animals interacting, even if it is tricky to demonstrate them beyond doubt. But one of the most awkward features about observational research is that the same situation is never repeated in exactly the same way: all sorts of variables are left uncontrolled. A merit of the experimental approach, to which we shall now turn, is that it allows much greater control, so that we can expect more consistent results. One can also set out to test a specific hypothesis rather than hoping that nature will come up with answers unaided.

1.4   The experimental approach

As will be apparent from the amounts of space I have devoted to the two topics, messages are rather easier to study by observation alone than are meanings. In some ways the opposite is the case when it comes to experiments. Placing animals in various different situations can give some idea of what causes them to produce signals, but many signals are only produced in the presence of other individuals and the behaviour of companions is not an easy factor to control. Because of this, it is much more fruitful to concentrate on one animal and to stimulate it with signals from models, recordings or simulations to see how it responds and thus to determine what meaning they have for it. Of course, if some of its responses are signals one may, incidentally, discover aspects of the message that they convey. One of the most fascinating things about communication is the way in which the behaviour of communicating animals interrelates; it is also unfortunately one of the main reasons why it is so difficult to separate out exactly what are the causes and functions of each of the acts that they perform.

In discussing experiments on the responses of animals to signals it is best to consider studies on auditory, olfactory and visual stimuli separately, as the different modalities present rather different problems. The most extensive work has been done on auditory stimuli and these are, in some ways, the easiest to study experimentally.

1.4.1   Auditory stimuli

Given the sophisticated recording and reproducing equipment now available, animal sounds can be played back with high precision to individuals. Experiments which involve making recordings of natural sounds and seeing how animals respond to them in the wild have most often been carried out on birds, usually to test the reaction of territorial males to the recorded song of other individuals (e.g. Weeden & Falls 1959; Goldman 1973; Wunderle 1978). If the loudspeaker is placed within the territorial boundary, males of most species will approach and fly up and down over it, or perch close to it and call (e.g. Harris & Lemon 1974; Martin 1980; Slater 1981). Some species tend to sing in response, whereas others do not. Experiments of this sort have shown that birds usually respond more strongly to the songs of strangers than to the songs of their neighbours (see Fig. 1.6), although a neighbour may elicit a strong response if the speaker is far from the appropriate territory boundary (Falls & Brooks 1975; Wunderle 1978). Responses to other species or to members of the same species from a totally different dialect area have sometimes been found to be comparatively weak (Harris & Lemon 1974) but may be strong when interspecific territoriality exists (Reed 1982). In general, however, an unknown intruder of the same species, with songs similar to those found in the neighbourhood, is usually the strongest stimulus.

These experiments give some idea of the meaning of song to territorial males, but there are also some difficulties of interpretation. Whereas in some cases the speaker has been placed close to the territorial boundary or outside the territory, so simulating the

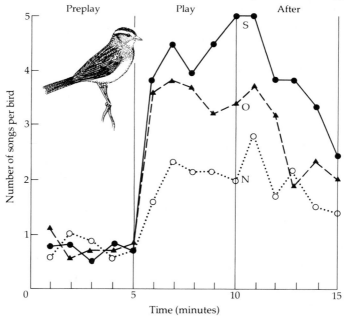

Fig. 1.6. The response of territorial male white-throated sparrows (*Zonotrichia albicollis*) to playback of stranger song (S), their own song (O) and the song of a neighbour (N). The song was played once every 15 seconds during the middle 5 minutes of a 15 minute observation period and the number of songs of the territory holder were scored for each minute. (From Brooks & Falls 1975b.)

song of a neighbour, in many it has been well within the boundary. The stimulus is therefore the song of another bird which is clearly on the bird's own territory but which cannot be discovered and which persists in singing despite responses by the territory owner. It is perhaps not surprising that territory owners are highly disturbed by this! Further difficulties with this approach lie in scoring responses when different birds may respond in quite different ways, in equating the output of the loudspeaker with the intensity which would be expected of a bird singing at that position (a point not often mentioned), in allowing for the fact that the signal will have been degraded by its passage through the environment before recording (Richards 1981a; see Chapter 3), and in the likelihood that minor differences in the signals used but not considered by the experimenter may be of relevance. This last point is most important where song is not very stereotyped but shows variations from place to place or between individuals. The response of a territory holder may depend a great deal on earlier experience of that particular song type or individual. All of these points have to be considered carefully, but perhaps the major point about using playback experiments to understand the meaning of signals, as with many other experimental techniques, is that it removes the interactive component from communication. Normally, two communicating animals influence each other and the response of one individual modifies the behaviour of the other. This is not the case in most playback experiments: the song is repeated a number of times or for a fixed period and the responses of the bird are noted, but the playback is not modified in the light of these. With greater understanding of communicative processes and with modern programming equipment there is obviously great scope for experiments to become more sophisticated by including some interaction, the course of the playback being influenced by the response of the subject.

While simple playback experiments have been a common approach in studies of bird song, some other experimental manipulations have been more illuminating. In a longer term experiment, Krebs (1976) showed that territories from which great tit (*Parus major*) males had been removed were less rapidly occupied by others if songs were played from loudspeakers located on them than if they were silent or the sound of a tin whistle was broadcast (Fig. 1.7). He later showed that loudspeakers playing a repertoire of several different song types were more effective than those playing only a single type (Krebs *et al.* 1978). These experiments added weight to the idea that song is an effective 'keep out' signal to other males, in other words that its meaning to them is

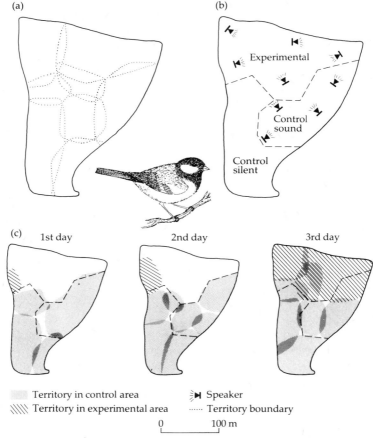

Fig. 1.7. Song as a male repellant in the great tit (*Parus major*). All the males were removed from the eight territories shown in (a). Three were then occupied by loudspeakers playing great tit song, two with speakers playing a similar phrase on a tin whistle, and three were left silent (b). Males reoccupied the wood as shown in (c), the three diagrams being after 8 hours, 12 hours and 20 hours of daylight. Although the whole wood was eventually recolonised, the area in which great tit song was played was occupied more slowly than the rest of wood. (From Krebs 1977.)

'Do not intrude', and that possessing a repertoire makes it particularly effective in this respect. The evidence for song attracting females is much less direct. In some species, singing declines after pairing (Catchpole 1973; Nice 1943), and in some it shows a resurgence if the female is removed (Krebs *et al.* 1981; Wasserman 1977). This is circumstantial evidence only; however, there is firmer evidence that, once attracted, females are affected by song. In some species, playback of song to captive females can lead them to show the soliciting posture normally shown by a receptive female in the presence of the male (West *et al.* 1979; Searcy & Marler 1981), suggesting that his songs are an important component of the displays inducing her receptivity. Song playback has also been found to influence the reproductive system of the female, leading to growth of the ova and ovulation, in budgerigars (*Melopsittacus undulatus*) (Brockway 1965) and canaries (*Serinus canaria*) (Kroodsma 1976) kept otherwise isolated from males.

Experiments have therefore shown that the song of male birds can have several different functions. All of these are achieved by the influence that it has on the responses of other individuals, from which in turn the various different meanings that song has for these individuals may be inferred. These studies have all used tape-recordings of natural song, although sometimes exploiting variations within its spectrum to test how animals respond to one song rather than another. A different experimental approach involves using sounds which have been synthesised or artificially altered to determine which aspects of the sound are important for its communicative function. Where it has several different functions these may be encoded in different parts of the sound. Several studies have been carried out on bird song to examine such questions (e.g. Brooks & Falls 1975b; Emlen 1972), but some of the best examples come from other taxa.

During courtship, male fruit flies (*Drosophila*) produce sounds with their wings which stimulate the female. Removal of the wings of males leads to greatly reduced mating success in *D. melanogaster*. Some improvement is achieved if an air current, such as would be produced by the wing movements of an intact male, is passed through the mating chamber, but the best results are obtained if this is supplemented by playback of the sound of normal wing movements (Bennet-Clark & Ewing 1967). The 'songs' produced by the wings consist of regularly repeated pulses of sound. The

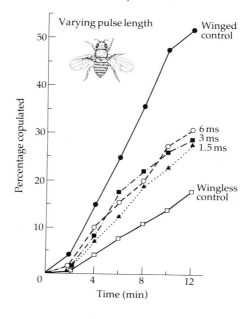

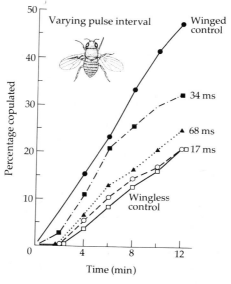

frequency of these sound pulses and the intervals between them vary from one species to another but are constant within a species, suggesting that they allow species recognition on the part of the female (Ewing & Bennet-Clarke 1968). Using simulated songs and wingless males, Bennet-Clark & Ewing (1969) were able to show that varying the pulse length made little difference to female receptivity, whereas this was reduced where the interval between pulses was greater or less than normal (Fig. 1.8). They concluded that the species-specificity is encoded in the intervals rather than in the pulse length.

Compared with bird songs, the sounds with which male amphibians attract females are also relatively simple and thus well suited to synthesis and modification (Capranica 1966). For example, Gerhardt (1974) has found female green treefrogs (*Hyla cinerea*) to be attracted as effectively by artificial calls as by natural ones, provided the former match up to certain criteria. Like the natural call, they must have roughly equal energy concentrated in two frequency bands with peaks at around 1 and 3 kHz; effectiveness is lowered if sounds at intermediate frequency are added. These experiments thus show how crucial the bimodal peaks are to mate attraction in this species.

These examples illustrate a number of ways in which auditory communication can be subjected to experimental analysis. There are some difficulties, but these are not nearly so great as those involved in experiments on visual and olfactory signals.

1.4.2   Olfactory stimuli

Humans are relatively insensitive to smell and are less dependent upon it than on visual and auditory stimuli; it is not, for example, easy to express olfactory sensations in words. For research on smells this has the advantage that it is easier to keep out subjective

---

Fig. 1.8. The effect of simulated song on mating in *Drosophila melanogaster*. The song of this species is produced by wing vibrations and is important in leading to copulation: in both diagrams more animals in the winged control group mate than do wingless controls. Wingless animals are more successful if courtship song is played over a loudspeaker. Variation in the pulse length of these sounds from the normal 3 ms does not affect their effectiveness (above), but doubling or halving the interval between pulses from the usual 34 ms makes the sound ineffective (below). (From Bennet-Clark & Ewing 1969.)

impressions than it is with the other modalities; but it has the disadvantage that it is not easy to know what questions to ask.

Because of their economic importance, a great deal of research has been devoted to insect sex attractants, the odours produced by females which attract males over a distance (Jacobson 1972). Many such substances have been isolated, analysed and synthesised. Either the isolated compound or its synthetic counterpart can be used to test attractiveness experimentally and generally it has been found, as might be expected, that these substances are species specific, tending to stimulate males of the same species much more strongly than males of other species (Shorey 1977). Pheromones with a wide range of other functions have also been isolated in invertebrates, and experiments on fish have shown pheromones which convey information on the identity of an individual, others which indicate its status, others which are produced by grouped fish and can subdue the aggression of territorial ones (Bardach & Todd 1970), and yet others which stimulate an alarm reaction (Smith 1976). All these different functions have been demonstrated quite simply by exposing fish to water drawn from tanks containing other fish, without the substances involved having yet been isolated. It is also true of mammals that extensive experimental work has been carried out on the influence of pheromones, particularly in the urine but also in vaginal secretions and from scent glands in various other parts of the body, without it being known exactly what substances are involved. At the most general level, possible influences of the soiled bedding or sawdust from the cage floor of one animal on the behaviour of another can be examined. If an effect is found then successive experiments with different secretions and fractions of them can pin down more exactly the origin of the substance responsible and its nature.

One advantage of experimenting on the effects of olfactory stimuli on animals is that they can be presented in a rather natural way. Signals which act as mate attractants travel over long distances and those involved in scent marking are left behind while the signaller moves on. In neither case is the presence of a signal necessarily associated with the presence of an animal in the wild, as it is with most visual stimuli. Thus natural reactions to smells can be more easily assessed, by measuring their attractiveness, the interest shown in them and various other changes in behaviour. As smells are persistent and signals based on them cannot be changed rapidly, the fact that experiments remove the interactive component from communication is of little import-ance: an animal which responds to a scent-mark by itself marking would be unlikely to expect a 'reply' until well after the experiment was over!

1.4.3 Visual stimuli

Experimenting on visual signals is more difficult than on auditory or olfactory ones because, except for a very small minority of cases, the receiver responds to the sight of the signaller itself showing a particular posture or display. This is not an easy stimulus to simulate without having a second animal present, although the increasing use of video equipment in behavioural research will perhaps lead to techniques involving presentation of visual signals just as auditory ones are presented using ordinary tape-recorders.

The influence of visual signals has been examined in several different ways. One of these, which has also involved the study of auditory components in signalling, has been the use of brain stimulation. Squirrel monkeys (*Saimiri Sciureus*) have had elec-trodes implanted in positions within the brain from which a particular signal could be elicited dependably by stimulation (Maurus & Ploog 1971; Maurus & Pruscha 1972). The animal was allowed to move freely within its group but could be stimulated at will by radio. Careful observation of the responses of other group members following a number of instances of a particular display then allows determination of the effect the display has on them. This is potentially a very useful procedure but it does have snags, largely connected with the interactive nature of normal com-munication and with the importance of context in determining the meaning of signals (Pruscha & Maurus 1976). The animal is pro-ducing a signal at the whim of the investigator rather than in response to a signal by a cage-mate or to a particular situation. If the context is an appropriate one, other individuals may respond as they would to the same signal produced naturally in that con-text. If the context is not right, then they may show surprise and alarm at the strange behaviour of their companion. Thus, the experimenter has to know a lot about the behaviour of his animals to choose the correct context; of course, if the context is exactly right the animal would presumably show the behaviour anyway!

Another way of presenting a live animal is by means of a mirror. This is a very strange animal—one which does the same thing as the subject at exactly the same time. Obviously this will not give an insight into sophisticated systems of communication but, at a gross level, the response may show how the animal reacts to seeing another at a particular location. Does it, for example, attempt to attack the mirror? The sequence of actions shown may also give some indication of how animals respond to each other. In the classic case where a male Siamese fighting fish (*Betta splendens*) is shown a mirror, he goes into full display and swims up and down alongside the apparent rival. He repeatedly moves to face the mirror but, as soon as he does so, he flicks back again to swim parallel to it. He is stuck in a loop between the 'facing' and 'broad-side' displays. Normally, the reaction to each of these is the other one (Simpson 1968), and two fish spend little time during a fight in the same posture as each other: thus an animal confronted by a mirror alternates frequently and continually between the two displays.

As this example shows, use of mirrors can tell us something about the causes of particular displays, but only in the most general terms because the 'companion' behaves in a totally in-appropriate way: animals do not normally mirror each other's actions. An alternative, which Simpson (1968) employed to find out the relationship between the facing and broadside displays mentioned above, is to devise puppets which are very like real animals and can be manipulated to show different postures. Simpson's puppet could be moved to face or be broadside to the fish he was studying and kept so for periods of time during which the displays shown by the subject were monitored. Halliday (1975) used a similar technique to study the courtship of newts, but with a live female held in a strait jacket as the equivalent of a puppet. This is useful here as the part played by the female newt in courtship consists largely of following the male or retreating from him, both actions which can easily be mimicked with the restrained animal, as they were for the experiments described in Fig. 1.9. Models have also helped to elucidate the meaning of some bird displays. In their studies of the glaucous-winged gull (*Larus glaucescens*), Stout and his co-workers used a variety of models ranging from crude wooden gull shapes, on to which stuffed head and wings could be mounted in various postures, to complete stuffed specimens

(Stout & Brass 1969). In more recent experiments radio-controlled, automated models have been used so that the experimenter has been able to alter the posture of the gull in the middle of the experiment. The models have been used to assess how gulls

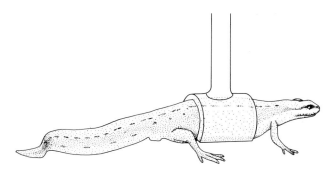

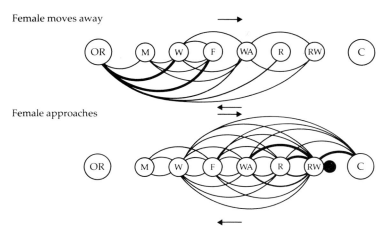

Fig. 1.9. The manipulation of courtship in the smooth newt (*Triturus vulgaris*) using a female in a strait jacket (above). In the flow diagrams the letters in circles indicate the sequence of male actions, from orientation (OR) through to creep (C), the thickness of the lines connecting them being proportional to the frequency of particular transitions. The transitions above are from left to right, forwards through the sequence, and those below from right to left. The sequence above is for transitions when the female has been made to move away from the male, that below when she is caused to approach. The effect of these on the male behaviour can be clearly seen. (From Halliday 1975.)

respond to different postures and so to determine what cues are important in the aggressive displays being studied. For example, static models with head raised were more likely to be attacked than ones with head lowered, suggesting that the latter posture inhibits attack (Galusha & Stout 1977). Experiments with the remote-controlled model showed that it was attacked more when it was turned from facing the gull to a position at right-angles to it, and less when rotated to face in the opposite direction, thus demonstrating how important orientation is in aggressive communication in this species (Hayward *et al.* 1977).

Where patches of colour or patterning comprise an important part of a signal these can be changed, either on a real animal or on a puppet, and the effect of this noted. Smith (1972) carried out such an experiment on wild male red-winged blackbirds (*Agelaius phoeniceus*) on their territories. These birds are largely black but with prominent red epaulets on their wings. When the epaulets of territory holders were blackened with a permanent dye, a high proportion of them lost their territories. However, those that succeeded in retaining them, but were unmated at the time of treatment, were still able to find and keep mates. On the basis of these results Smith suggests that the epaulets are primarily a territorial signal between males with little if any influence on females. A similar manipulation was carried out by Rohwer (1977) on Harris' sparrows. In this species, dominant individuals have considerably more black feathering on the throat and crown than subordinates, and Rohwer (1975) argued that this acts as a signal of their status. However, the signals alone could not confer status: subordinate birds with throats and crowns dyed black were persecuted by legitimate dominants, and dominants with these feather areas bleached did not become subordinate but had to engage in many fights to maintain their dominance. Later experiments by Rohwer and Rohwer (1978) suggested that these results were obtained because the animals were showing behaviour which was incompatible with the status signals they had been given artificially. Subordinates given testosterone as well as being dyed increased their social status without being persecuted.

Rather than applying paint or dye to individual animals, it is perhaps more realistic to bring them into a condition whereby they show a changed signal naturally. Rhesus monkey females are most attractive to males in the middle of their menstrual cycles, at around the time of ovulation, and at this time the skin round their vaginal opening is bright red in colour. Michael and Keverne (1970) were able to show that the attractiveness does not depend on this colouration. Using females whose ovaries had been removed, and were therefore unstimulating to males, they made the sex skin of these animals red by rubbing the hormone oestrogen on to it. Despite this, males still found these females unattractive. The measure of attractiveness used was an interesting one. The males had to press a lever 250 times to gain access to a female: they would do this for a normal female, but the labour involved was too great when the reward was an ovariectomised female or one which merely had red sex skin. By further experiments Michael and Keverne were able to show that the signal which attracted males was a pheromone produced from within the vagina itself.

These examples, drawn from the three main sensory modalities, give some idea of the problems with, and potentialities for, carrying out experiments on signalling systems which make use of each. Many different approaches, involving a great deal of ingenuity, have been used and some impressive findings have accumulated. But it is early days yet. With the solid foundation of observation and simple experiment which has been laid, it should not be too long before interactive experiments are a commonplace, the observer modifying the stimulus in ways appropriate to the response shown by the subject.

1.5   Conclusion

This chapter has been primarily concerned with methods and, because a great many different approaches have been used in the study of communication, each of these has had to be touched on rather briefly. The aim has been to indicate how interactions between animals can be broken down into component parts and the influence of one animal on the other determined. At first sight discovering what causes an animal to produce a signal and what effect this signal has on another might seem a simple task. In fact it is far from being so. The speed with which one display can follow another and the fact that communication is mutual, interacting animals continually altering their behaviour in the light of their partner's actions, make it hard to determine the effect of each signal from observation alone. Experimental manipulations may simplify matters but, by removing the interactive component, they may place the animal in an artificial situation in which its responses have little to do with normal communication. But, taken together, the sophisticated quantitative methods available for analysing observational data and the ingenious experimental manipulations which have been devised can give us some insight into both the causes and functions of different signals.

1.6   Selected reading

Burghardt (1970), Hinde (1975, Chapter 5) and MacKay (1972) all provide useful discussions of what it meant by communication. The article by Smith (1968) is a good introduction to his ideas on message/meaning analysis, and Cullen (1972) also gives insight into the problems of studying communication. The analysis of sequences of behaviour is discussed by Slater (1973), and Dingle (1972) gives a clear introduction to the application of information theory in studies of this sort. Although it does not concentrate on methodological issues, the book edited by Sebeok (1977) is certainly the major reference book in the field; it includes chapters dealing with the use of the different sensory modalities as well as on communication in different animal groups. The most thorough up-to-date review is that by Green and Marler (1979).

CHAPTER 2
INFORMATION
AND
COMMUNICATION

TIM HALLIDAY

2.1   Introduction

In the previous chapter, a distinction was made between the *message* encoded in an animal signal, and its *meaning*. The message is a function of the state of the individual that produces the signal; the meaning is the information extracted from the signal by an animal that perceives and responds to it. Animals convey many kinds of message; that is, they communicate with one another about many different things. The aims of this chapter are, first, to categorise the various kinds of information that animals communicate to one another and, secondly, to relate certain properties of signals to the nature of the messages that they contain.

Examples will be given of signals performed by a variety of animals and perceived through a variety of senses, including vision, hearing and olfaction.

2.2   Some general principles

We may consider the nature of animal signals as if they have been 'designed' for a specific purpose. This is a shorthand way of saying that we assume that the signals we observe are the product of natural selection, which has favoured those properties of signals that make them most effective at conveying information.

A basic design feature of any signal is how specific is the message that it conveys. This will be related to how much variation there is in the performance of a signal or, in other words, how stereotyped the signal is. A signal that is absolutely constant in form can only convey a very simple message. Thus, members of a species that perform a single, completely stereotyped threat display can only indicate that they are aggressive; they cannot convey *how* aggressive they are. However, the stereotyped nature of their threat display will tend to ensure that it is readily and unambiguously recognised. Conversely, a signal that is variable in form can, potentially, communicate subtle variations of a basic message, but if it is very variable there is a risk that a perceiver may confuse it with another signal conveying a different message. Furthermore, in an environment in which a signal is performed against a high level of background 'noise', a receiving animal may be unable to differentiate one variant of the signal from another.

For any given signal, we may suppose that there is some optimum level of stereotypy. This will depend on a variety of factors, including the following.

(i) The nature of the message contained in the signal. Signals, or components of signals, that indicate the species of the performer convey unequivocal information and tend to be highly stereotyped (see section 2.3). Conversely, signals whose message relates to the aggressive or sexual motivational state of an animal may be expected to be variable, since the information they convey is intrinsically changeable (see section 2.8).

(ii) The capacity of potential receivers of the signal to differentiate between variants of the signal. This will depend both on the capacity of the relevant sense organ to detect small variations in signals and on the extent to which the receiver's nervous system can process complex sensory information.

(iii) The susceptibility of the signal to environmental influences. We would expect signals to be more stereotyped if they are performed in conditions that may cause them to be distorted or easily confused with other sensory input to the receiver.

Until quite recently, ethologists tended to emphasise the stereotyped nature of animal displays. Lorenz's (1937) assertion that many behaviour patterns, especially certain displays, are as characteristic of a species as features of its anatomy was a crucial step in the foundation of ethology. However, research in the last few years has tended to focus on variation in the form of animal signals and the extent to which this conveys information that can be detected and responded to by a receiver. For example, male toads (*Bufo bufo*) produce a very simple characteristic call when fighting for a female. Detailed analysis reveals that there is important variation in this call. Large males have slightly 'deeper' voices than small ones and, as a result, each male is able to vary how persistently he fights, depending on his opponent's size, which is assessed from the pitch of its call (see page 58).

Behaviour patterns that function as signals, referred to as *displays,* are presumed to have evolved from actions which ancestrally fulfilled quite different functions. For example, the sham-preening display of the mandarin duck (Fig. 2.1) appears to have evolved from a preening movement. The evolutionary changes that such a movement underwent between its ancestral and its present forms are called *ritualisation.* During ritualisation, signals may become more exaggerated, more elaborate, associated with bright colours or special structures such as plumes or crests, as well as more stereotyped. Hazlett (1972) analysed the aggressive claw-waving display of a crab, *Microphrys bicornutus,* and the movement, used during feeding, from which the display has apparently evolved, and compared their variabilities in form (Fig. 2.2). During the display, the crab's claw tends to be raised through a higher and rather less variable angle than during feeding.

Because movements like the crab's claw-waving display are recognisable to human observers as distinct patterns of behaviour, early ethologists tended to categorise them as what Lorenz (1937) called *fixed action patterns (FAPs)* and Morris (1957) pointed out that many displays are performed with a 'typical intensity'. In the light of research that has shown displays to be quite variable in form, Barlow (1977) has suggested the use of the more realistic term *modal action pattern (MAP),* which implies that movements show statistical, rather than absolute, stereotypy. At each individual performance, a display may depart more or less from its modal form. The emphasis of much contemporary research into animal communication is on investigation of the extent to which

Fig. 2.1. The sham-preening display of the male mandarin duck (*Aix galericulata*).

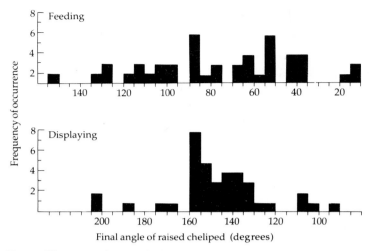

Fig. 2.2. Histograms showing the frequency of occurrence of different angles of the final position at which the cheliped is held by the spider crab *Microphrys bicornutus*. Above: data from feeding crabs. Below: data from crabs in aggressive interactions. (From Hazlett 1972.)

this variation is itself an important means of conveying information from one animal to another.

2.2.1   *Sensory perception of signals*

Animal signals are perceived by sense organs and, in some species, the highly specific nature of a signal is paralleled by the specific responsiveness of the sense organ that perceives it. The female silk moth *Bombyx mori* produces an airborne sexual attractant, called bombykol, and males possess huge antennae containing millions of receptor cells that are specifically sensitive to this substance (Schneider 1974). The ears of most female frogs contain very large numbers of sensory cells specifically sensitive to the sound frequencies that predominate in the advertisement calls of conspecific males, but relatively few that are sensitive to other frequencies. The male Puerto Rican treefrog *Eleutherodactylus coqui* produces a biphasic call, consisting of a low-frequency 'co' and a high-frequency 'qui'. The ears of males contain large numbers of cells tuned to the 'co' frequency but are barely sensitive to the 'qui'. In contrast, females have ears that are especially sensitive to the higher pitch of the 'qui' (Narins & Capranica 1980; section 4.5.1).

Among birds and mammals, sense organs are generally not as specifically adapted for perceiving stimuli of biological relevance as the antennae of silk moths or the ears of frogs. Their eyes and ears are typically able to perceive light and sound over a wide frequency range. Thus, for birds and mammals, the discrimination between one signal and another is largely carried out in the central nervous system, whereas for a frog the ability to discriminate is 'hard-wired' into its sense organs.

2.2.2   Complex signals

It would be misleading to imply that each signal that an animal produces conveys a single, specific message. The 'co-qui' frog call contains two messages, each perceived by a different category of perceiver. To males, the 'co' is an aggressive signal which is involved in their being widely spaced. To females, the 'qui' is a sexual signal which enables them to locate males.

Nuechterlein (1981) has analysed the advertisement call produced during the breeding season by both sexes of the western grebe (*Aechmophorus occidentalis*). Individuals of this species belong to one of two colour phases, light and dark, that differ in bill colour and in the extent of dark colouration on the crown. Variations in the form of the advertisement call can be related to the individual identity, sex, colour-phase and pairing-status of birds, as well as to their geographical location and to the stage of the season. Advertisement calls performed by males attract females, who may respond by calling back. Birds that have formed pairs continue to use the call to maintain contact with one another. Single males ignore the calls of other males and of females who are already paired, but respond to the calls of single females, especially if they belong to the same colour phase as themselves. The most important variable in the call is the length of the bouts in which it is performed. Paired birds produce shorter bouts of advertisement calls than single birds. Paired birds also alternate their calls with those of their mates. As a result, the calls of single birds can be differentiated by other birds because they tend to be performed in isolation from other calls. Thus, what at first appears to be a simple signal conveys a variety of messages.

2.3   Signalling species identity

Mating with an individual of a different species can be one of the most maladaptive things that an animal can do. Indeed, in a species which reproduces only once during a lifetime it is equivalent to being eaten by a predator, since the individual's entire reproductive potential will usually be lost as a result of a hybrid mating. A major function of many of the displays that precede mating in animals is to ensure that individuals recognise, and mate exclusively with, members of their own species.

Signals which enable potential mates to locate one another are typically highly species-specific. Natural selection has clearly favoured those aspects of signals that make them more distinct from the signals of other species with which hybrid matings might occur. Male fireflies (*Photinus*) signal to females, who sit among foliage, by producing flashes of light as they fly around at night (Lloyd 1966, 1971; section 3.3.6). Each species has a characteristic pattern of male flashes (Fig. 2.3), and females respond only to the male flash pattern that is appropriate to their particular species. For many frogs, the risk of engaging in hybrid matings is increased by the fact that, in some parts of the world, several species share the same breeding sites and the same breeding season. Littlejohn (1977) has described how a number of Australian frogs minimise the incidence of hybrid matings by a combination of spatial segregation and mating call differences (Fig. 2.4). The most

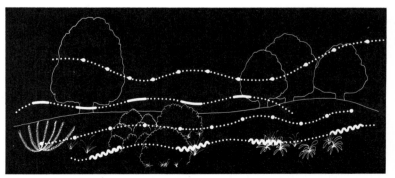

Fig. 2.3. Sexual display flashes and flight paths of male fireflies of four different species, as they would appear in a time-lapse photograph. Species (from top): *Photinus brimleyi*, *P. collustrans*, *P. ignitus* and *P. granulatus*. (From Lloyd 1966.)

marked differences between species' mating calls are shown by those species that mate in the same part of the pond, for which the risk of hybrid mating is greatest. Among insects, male antlions are unusual in that it is they, rather than females, that advertise themselves to prospective mates. From their large thoracic glands they produce a volatile secretion that contains two chemical com-

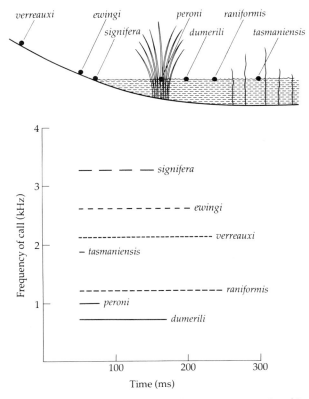

Fig. 2.4. Spatial and acoustic mating isolation between seven species of Australian frogs that breed at the same time and in the same pond. The species are: *Limnodynastes dumerili, peroni* and *tasmaniensis, Litoria ewingi, raniformis* and *verreauxi* and *Ranidella signifera*. Upper figure: positions in the pond from which males tend to call. Lower figure: diagrammatic representation of the principal sound frequencies and temporal patterning of notes in the male calls. Note that whereas species that call close together (e.g. *ewingi* and *signifera*) have very different calls, species that call far apart (e.g. *verreauxi* and *raniformis*) have rather similar calls. (From Littlejohn 1977.)

pounds (Löfquist & Bergström 1980). Among three species living sympatrically in Sweden, the male sexual attractant of each species contains a unique 'code' of two compounds (Table 2.1), though two of the three species have one of the compounds in common.

Table 2.1. The chemical composition of the male sex-attractant secretions of three species of antlion found sympatrically in Sweden. (From Löfquist & Bergström 1980.)

Species	Molecular weights of the two principal compounds in the male's sex-attractant	
Euroleon nostras	152	170
Grocus bore	166	170
Myrmeleon formicarius	154	168

In many animals, selection favouring divergence of the sexual display of one species from that of a closely related species has produced a situation in which differences in behaviour are much the most marked differences between them. The morphologically identical American treefrogs *Hyla versicolor* and *H. chrysoscelis* were not recognised as being distinct species until the marked differences in their male advertisement calls were discovered (Johnson 1959; see also section 3.6.1). They are good species in the sense that hybrids between them have very low viability. Likewise, a number of cryptic species of Hawaiian *Drosophila* are differentiated solely on the basis of their sexual displays (Carson 1982).

The songs of many birds are very elaborate and certain aspects of them show a lot of variability, both within and between individuals. By playing normal and modified recordings of con-specific songs to male indigo buntings (*Passerina cyanea*), Emlen (1972) was able to pinpoint those components of male song that are important in species recognition. Whereas the order in which different notes within the song were heard by males did not effect their response, changes in the time intervals between notes and in the sound characteristics of the notes themselves did markedly reduce the ability of males to recognise songs as belonging to their own species. Surveying a number of passerine species, Emlen concludes that species recognition is based on those features of song that show the least variation.

Whereas, as described in section 2.2.1, the ability to respond selectively to conspecific courtship displays is 'hard-wired' into the nervous systems of some animals, individuals of other species acquire this ability by learning. For example, some birds learn the characteristics of their species' song by listening during their early development to the songs of adult conspecifics.

2.3.1 Species differences and other selective forces

Despite the fact that the risk of hybrid matings will generate powerful selection favouring the divergence of the sexual signals of closely related species, differences between the signals of some species seem to be far greater than one would think necessary to ensure reproductive isolation between them. For example, the characteristic calls of song-birds, that enable humans readily to distinguish one species from another, seem to show more differences between species than the function of species recognition should require, particularly in view of the excellent ears possessed by birds.

Recent studies of Darwin's finches (*Geospiza*) have shown that, although the songs of different species cannot reliably be differentiated from sound spectrograms, males are able to discriminate between conspecific and heterospecific songs (Ratcliffe, pers. comm.). The lack of variation between these species may be related to the fact that they are virtually the only passerines on the Galapagos, where the habitat is very sparse and open. In more heavily vegetated environments, where there is much greater diversity of song-birds, selection for species-specificity may be much more intense because the acoustic environment is more complex.

Much of the elaboration of bird song may be attributable to sexual selection rather than to selection for species distinctiveness. Males of many species have to compete with one another for mates, either directly, through fighting, or indirectly, through differences in their degree of attractiveness to females. If females prefer those conspecific males with the most elaborate songs, then there will be powerful selection favouring increasing complexity of male song. Thus, relatively small differences between the displays of different species may become greatly exaggerated by sexual selection (Halliday 1978).

2.4   Signalling group identity

In animals that form stable groups it may be important for individuals to be able to recognise other members of their particular group. All the bees belonging to the same swarm share a distinctive odour and will attack individuals that carry a different smell (Wilson 1971). African dwarf mongooses (*Helogale undulata*) live in well-defined social groups and individuals frequently mark fellow group members with secretions from their cheek and anal glands (Rasa 1973). This behaviour is especially important among juveniles and appears to be a social bonding mechanism by which animals become accepted as members of a group. Colony odours are recognised by rats, but may be due to accumulated faeces and urine as much as to odours produced by colony members (Brown 1979).

Many territorial song-birds, though they do not live in coherent social groups, are able to recognise and respond differentially to individuals who come from the same locality as themselves. In the Indian hill mynah (*Gracula religiosa*), both males and females produce a variety of calls in addition to a number of basic vocalisations common to all individuals (Bertram 1970). Each bird, male or female, has a repertoire of between five and twelve calls, none of which it shares with its mate, though it has several in common with neighbouring individuals of the same sex. As a result of each bird learning calls from its neighbours early in life, there are pronounced local dialects in call repertoires, and it is only at distances greater than about 14 km that two individuals of the same sex will have no calls in common. In mynah birds, individuals respond more strongly, in terms of calling themselves, to calls with which they are familiar than to strange calls. This contrasts with the behaviour of many territorial male song-birds, such as the white-throated sparrow (*Zonotrichia albicollis*) (Brooks & Falls 1975a), which respond more strongly to strange songs than to the songs of their neighbours.

An essential component of recognition of the group to which an individual belongs is learning. Through familiarity with the appearance, odour or vocalisations of their social companions or territorial neighbours, animals become able to discriminate between strangers and non-strangers. Often this discrimination may be based, not on recognition of an individual simply as a member of a group, but on recognition of it as an individual. This is discussed in the next section.

2.5 Signalling individual identity

In the social lives of many animals it is important that they respond in particular ways to particular individuals. For example, they may behave sexually to their mates and aggressively to others, or be less aggressive to territorial neighbours than they are to strangers. In order that they may vary their behaviour in this way, they must be able to recognise individuals. In many species it is likely that individual recognition is based, largely or wholly, on variations in appearance, odour or voice quality. Individual recognition on the basis of such variations must involve learning and is only likely to occur in social animals that have the opportunity to become familiar with one another.

Dwarf mongooses scent-mark objects, and one another, with secretions from their cheek and anal glands (Rasa 1973). The cheek gland secretion is a form of threat and elicits reciprocal marking in an animal that smells it. Mongooses do not differentiate between the cheek secretions of different individuals. In contrast, the anal gland secretion is not a threat signal, but does elicit varying responses depending on the identities of the marker and the perceiver. It also provides information about the time at which a scent mark was made; mongooses can differentiate between marks made an hour apart. Individual recognition in another species of mongoose, *Herpestes auropunctatus,* is thought to be the result of each animal producing a limited number of volatile compounds in different relative amounts (Gorman 1976).

Brooks and Falls (1975a, b) have investigated the basis of individual recognition by song in male white-throated sparrows, which discriminate between the songs of neighbours and strangers. They found that what was important was not the overall pattern of notes within a song, but certain characteristics of the song, particularly of the first three notes. Using playback of experimentally manipulated song recordings (see Chapter 1), they found that changes greater than five to ten per cent in the pitch of the whole song, or of just the first note, were sufficient significantly to reduce a male's ability to recognise a neighbour's song. In contrast, changes as large as fifteen per cent in the duration of the whole song, or of the first note alone, had no effect on neighbour recognition.

The songs of birds typically contain a number of features that show little variation within and between individuals, and others that vary a lot. In most birds, songs are developed by imitation of conspecifics, and variations between individuals are sometimes attributed to 'copying errors'. However, since such variations can be important in individual recognition, we may regard the ability to produce songs that are slightly different from those heard during development as an adaptive feature in certain species. Of course, to label the differences between the songs of different individuals as 'copying errors' is to make the implicit assumption that birds 'try' to learn songs accurately, which is essentially untestable.

Individual recognition is especially important in the relationship between a parent and its young. For parents, it is vital that they care only for their own progeny. For young animals, recognition of their parents may be essential for their survival; adult herring gulls (*Larus argentatus*), for example, readily devour chicks that are not their own. In most precocial animals, the primary initiative in the formation of parent–young bonds comes from the young, who rapidly learn the characteristics of their parents during a sensitive period very early in life (Bateson 1966, 1973). Parental recognition of their young may take longer.

Reliable parent–young recognition is particularly important in dense breeding colonies, such as that of the common tern (*Sterna hirundo*). Adult terns catch fish out at sea and bring these back to feed their chicks. As they come in to land, they give a characteristic call, and common tern chicks are able to discriminate between the feeding calls of their parents and those of other adults (Stevenson *et al.* 1970). When they hear a parental call, chicks call in response, and the parent is able to land next to its own chick. There is a good deal of evidence among birds for recognition by voice between parents and chicks (Beer 1970). The extent to which chicks can recognise parents and the age at which they become able to do so vary between species and depend on ecological factors. Recognition of parents tends to be more marked and more rapidly developed in species that nest in very dense colonies and in species in which the young leave the nest at an early age; these are the conditions under which chicks are most likely to get lost. Chicks of the kittiwake (*Rissa tridactyla*) never develop this ability. Since the young do not leave the nest until a late stage, parents can always find them by returning to their nests (Beer 1970).

Individual recognition may also be important in the maintenance of a bond between breeding partners. Brooke (1978) has shown that female Manx shearwaters (*Puffinus puffinus*) are able to recognise the calls of their mates. In some species of birds, it appears that mate recognition is enhanced by individuals modifying their vocalisations in response to those of their mates. Paired male and female twites (*Acanthus flavirostris*) have very similar flight calls that differ markedly from those of their neighbours (Marler & Mundinger 1975). In some species, pairs perform duets in which partners either sing the same songs in unison or each contributes different components to a composite song (Thorpe 1972). Established pairs of parrots (*Trichoglossus*) perform not only synchronised songs but also a number of elaborate postural displays that they execute together (Serpell 1981). Individual recognition is only one of many possible functions of mutualistic displays between paired animals. Such behaviour patterns may also serve as threat signals directed at individuals outside the pair, or as a mechanism that maintains and strengthens the pair bond (Wickler 1980).

2.6   Signalling kinship

There are at least two very different reasons why it may be important for animals to recognise their close relatives. When mating, it is adaptive for most animals to avoid their close kin, so that their progeny do not incur the harmful genetic effects of inbreeding. However, for some species it may be adaptive to mate with relatives, though at the level of cousins rather than siblings (Bateson 1980), and Bateson (1982) has shown that Japanese quail (*Coturnix coturnix*) choose to mate with their cousins rather than with less or more closely related individuals. In a quite different context, altruistic behaviour, in which individuals confer a benefit on other animals at some cost to themselves, is often only explicable in evolutionary terms if it is directed towards kin. The more closely related is the beneficiary of an altruistic act, the more strongly is the act favoured by natural selection, or, more precisely, by kin selection.

It is likely that kin recognition is most commonly achieved through individual recognition. Since many animals associate closely with their immediate relatives during early life, they have ample opportunity to learn their individual characteristics. Female chimpanzees abruptly stop consorting with those males with whom they have previously associated when they first come into oestrus (Pusey 1980); their previous male companions are very likely to be their close relatives.

Tadpoles of the American toad (*Bufo americanus*) swim around in densely packed schools and, at least under laboratory conditions, show a preference for association with their siblings (Waldman & Adler 1979). It is possible that the mother contributes some substance to her progeny that enables them to recognise one another, but experience of their siblings also plays a role in the development of the tadpoles' preference to school with one another (Waldman 1981). In the social desert woodlouse (*Hemilepistus reaumuri*), parents can recognise their young, as many as 100 of them, on the basis of a family odour which results from the young exchanging secretions among themselves (Linsenmair 1972). Greenberg (1979) has shown that sweat bees (*Lasioglossum zephyrum*) learn to recognise their kin on the basis of genetically determined odours.

It is possible for some animals to recognise their kin even though they are not familiar with them. Bateson (1982) has shown that the preference of Japanese quail for mating with their cousins is expressed even if they have not encountered them before. It appears that they achieve this by having a general preference for birds slightly different in appearance to those with which they were reared and which typically will be their siblings. Wu *et al.* (1980) have shown that pigtail macaques (*Macaca nemestrina*) can recognise their kin even after they have been separated from them since birth.

In these examples the recognition of kin can only be called communication in a rather tenuous sense. These animals are not apparently emitting specific signals that indicate 'I am your brother' or 'I am your cousin'. It is unlikely that any such signal does exist among animals. However, if variations in the form of a particular signal can be used by perceivers to identify the degree of relatedness of a performer, then this may be an important functional component of that signal. For example, in many song-birds there are clear local 'dialects'. If, in a particular species, individuals do not disperse far from their natal area, birds that share a dialect will tend to be closely related, and so dialect could be used as a means of recognising kin, though there is no evidence for this.

A very important kind of kinship recognition is that between parents and their offspring, as discussed in section 2.5.

2.7   Signalling competitive status

The struggle for survival among members of a species provides one of the fundamental conditions under which natural selection acts. Individuals may compete for food, water, space, nest sites, mates and many other resources essential for their survival and reproduction. Competition may be a costly activity. At the very least, it takes up time and uses up energy; at worst, if it involves fighting, it may lead to injury or death. The costs incurred during competition diminish the value of the benefits gained from the resource that is being contested. Thus, we would expect natural selection to have favoured patterns of behaviour that enable animals to minimise the occurrence, duration or severity of competitive interactions. One way by which they could do this is to exchange information, early in a competitive encounter, about their respective strengths. An animal that perceives itself to be the weaker of two potential contestants should avoid a competitive encounter, since it will incur the costs involved in that encounter but is very unlikely to derive any benefit.

In many species, differences in competitive ability are signalled by readily perceived features of individuals, such as their overall body size or the size of any weapons that they possess. In some animals, individuals of high status may have characteristic features, such as the manes of adult male lions and baboons. Among male Harris' sparrows (*Zonotrichia querula*) an individual's rank in a dominance hierarchy is correlated with the amount of black colouration on his breast plumage (Rohwer & Rohwer 1978; section 1.4.3). In a few animals, competitive status is signalled by behaviour patterns, rather than by such anatomical features as size and colouration, and this appears to be related to conditions in which the appearance of a rival is not easily perceived. For example, male red deer (*Cervus elaphus*) establish territories during the rut (breeding season) and signal their territorial status by roaring. Using recorded roars, played to stags in the field, Clutton-Brock and Albon (1979) have shown that roars produced at a high rate inhibit aggression in stags that hear them. Thus many potential aggressive encounters between stags are probably avoided without rivals having to see each other at close quarters,

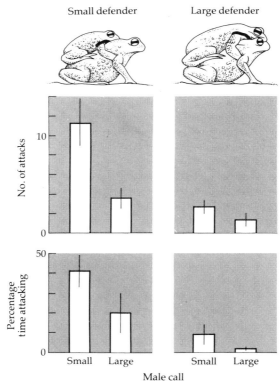

Fig. 2.5. Results of an experiment designed to test the hypothesis that the calls given by male toads (*Bufo bufo*) during fights are used by opponents to assess the relative size of their opponents. Medium-sized males were allowed to attack small or large males who were already in amplexus with females and who were prevented from croaking by rubber bands passed through their mouths. Each attacker was used in two experiments. In one, he heard tape-recorded croaks of a small male and in the other the croaks of a large male. One set of 12 males attacked small defenders and another 12 attacked large defenders. Two measures of attacking behaviour were used, the number of attacks and the percentage time spent attacking. The histograms represent mean scores, with vertical lines indicating the standard errors of those means. (From Davies & Halliday 1978.)

let alone engage in a fight. Male toads (*Bufo bufo*) fight for females, whom they outnumber in a breeding population, and it is generally the larger males who win (Davies & Halliday 1979). Much of this fighting occurs at night or in murky pond water, so that rivals may not readily be able to assess one another's size directly. As they fight, male toads produce soft, peeping calls and the frequency, or pitch, of these calls is inversely correlated with male body size, so that small males produce high-pitched calls, large males low-pitched calls. Using recordings of the calls of small and large males, Davies and Halliday (1978) have shown that a male toad fights less persistently if his rival produces calls of a lower pitch than his own (Fig. 2.5). The effect of toad calls is thus to shorten fights, rather than to avoid them altogether.

An important question relating to the signalling of competitive status is whether the signals involved are reliable, or whether animals enhance their true status by producing signals that convey false or inaccurate information. This issue is discussed in Chapter 5.

2.8   Signalling motivational state

In the course of their social lives, animals will interact with one another in many different ways. On some occasions they may be rivals for a resource such as food, on others they may be potential mating partners, and sometimes they may simply be attempting to settle close to one another. When one individual approaches another there is likely to be considerable uncertainty about how each will respond to the other. This uncertainty can be reduced if each produces signals that convey information about its internal state. In this section, we consider signals involved in just two kinds of social interaction, aggression and sex.

2.8.1   *Signalling aggression*

Many aggressive signals involve the prominent display of offensive weapons, such as teeth, claws or horns, that are actually or potentially used in fighting. Signals that indicate an intention or a readiness to fight are referred to as *threat* signals. Although some animals do at times engage in fights that result in serious injury or death, it is more often the case that aggressive encounters are resolved by an exchange of threat signals.

A common feature of threat signals is that they involve postural changes that increase the performer's size as perceived by a rival. For example, cats arch their backs, raise their fur and fluff up their tails, and Siamese fighting fish (*Betta splendens*) extend their fins and open the opercula that cover the gills. Threat displays may also incorporate colour changes, the production of sounds and the secretion of odours, depending on the species.

Change in apparent size during threat has been interpreted in two very different ways. Classically, it has been suggested that a threatening animal makes itself appear larger than it really is so as to intimidate its rival. An alternative argument is that by extending its body to a maximum extent, an animal indicates, not that it is larger than it acually is, but exactly how big it is. This is a question of whether displays are 'honest' or 'deceitful' and is discussed in Chapter 5.

Another important issue is the extent to which aggressive signals are stereotyped; whether, by being variable in form, they can convey information about an animal's level of aggressive motivation. It is important to stress that differences in the degree of stereotypy of different signals may be more apparent than real. There are major limitations in our ability as human observers to detect small variations in the performance of animal signals. A signal that appears to our senses to be very stereotyped may, when analysed in detail by means of cine film or sound spectrograph, prove to be highly variable. If we succeed in detecting and quantifying variation in animal signals, by means of appropriate technology, there remains the further important question of whether such variation is of any biological significance. This is a matter for empirical testing by appropriate experimental procedures (see Chapter 1).

Aggression in lizards

Recent studies of the threat displays used by certain species of lizard illustrate how complex what at first appears to be simple, ritualised behaviour can be.

The helmeted lizard (*Corythophanes hernandezii*) of Mexico has a threat display that incorporates many of the features that we classically associate with such a behaviour pattern (Fig. 2.6). In an aggressive interaction between two lizards, the contestants align themselves side by side, inflate dewlaps under their chins, darken a black pattern on the head and neck, and laterally compress their bodies. The prominent helmet on the top of the head does not change in size but the other elements of the display present the rival with an image of an animal that is considerably larger than one that is not threatening (Carpenter 1978). Such behaviour may be regarded as 'typical' lizard threat, but studies of other species reveal a more subtle system for communicating aggressive motivation. Jenssen (1977) has analysed the threat signals of anoline lizards and has shown that they consist of a basic display

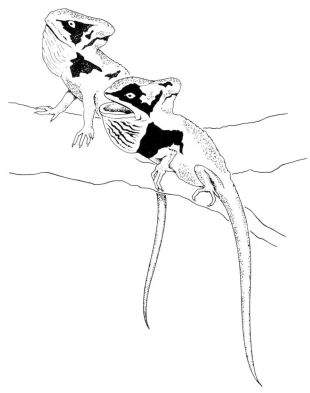

Fig. 2.6. Threat display of the Mexican helmeted lizard (*Corythophanes hernandezii*). Besides having a permanent 'helmet', a threatener's apparent size is increased by extending the dewlap and opening the mouth. In addition, the patterns on the helmet, dewlap and shoulder become darker. (From Carpenter 1978.)

that is performed in conjunction with one or more 'modifiers'. For example, in *Anolis opalinus*, Jenssen (1979) has described seven static and six dynamic modifiers that may be performed in association with a basic head-bobbing threat display.

Static modifiers consist of body postures and include erected crest, gorged throat and opened mouth. These features alter the appearance of the displaying lizard and can be performed in different combinations. Static modifiers can be ranked in order of

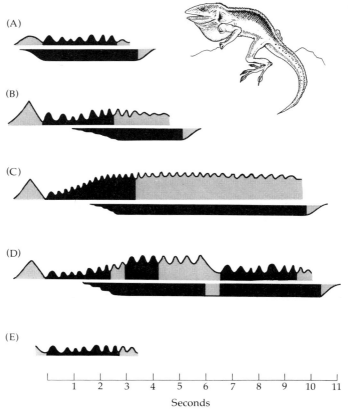

(A)

(B)

(C)

(D)

(E)

        1   2   3   4   5   6   7   8   9   10   11

                              Seconds

Fig. 2.7. Examples of five distinct head-bob threat displays used by the lizard *Anolis limifrons*. In each diagram, the upper portion shows the amplitude of vertical head movements, the lower portion the amplitude of dewlap extension. Black areas indicate components of the displays that are always present, grey areas those components that are sometimes present. (From Hover & Jenssen 1976.)

the frequency with which they are performed, and a combination of rare ones with commoner ones indicates a high level of aggression. Thus, a head-bob display that is combined with the commonly performed erected-crest posture indicates a lower level of aggression that one that is combined with both erected-crest and the rarer opened-mouth posture.

The six dynamic modifiers are movements of part or all of the body, and include rearing-up, pulsed inflation of the dewlap and rolling the head. The incorporation of dynamic modifiers into an animal's aggressive display depends on the context in which it is displaying. Whereas dewlap-pulsing appears in all aggressive contexts, rearing-up is performed only in interactions between males.

Jenssen (1977) concludes from his studies of anoline lizards that display modifiers may convey more information than the basic displays with which they are associated. While the basic display indicates aggressive motivation, the modifiers signal the level of that motivation.

Level of aggressive motivation is coded in a rather different way by the threat displays of the lizard *Anolis limifrons* (Hover & Jenssen 1976). This species performs five distinct patterns of head-bobbing during aggressive interactions, called patterns A to E (Fig. 2.7). Which of these displays is performed depends on the distance between two protagonists; as animals approach, their displays progress up the scale from A to E. This suggests that display E indicates the highest level of aggression and display A the lowest. When two lizards display to one another, the answering animal usually responds either with the same display or with the next highest in the A to E scale.

These examples emphasise the point that what at first may seem to be very stereotyped displays prove on close analysis to be very variable and to be capable of conveying quite complex information. Lizards are able to convey subtle 'shades of meaning' when behaving aggressively to one another.

2.8.2 Signalling submission

In an aggressive interaction between two animals, there will generally be a winner and a loser. If the potential costs of fighting are high, it will be of considerable benefit to an animal that is going to lose to bring the contest to an end quickly and in a way that minimises the risk of injury. In many species, animals perform displays or adopt postures the effect of which is to reduce the level of aggression shown towards them by a rival; these are called *submission* or *appeasement* displays.

Darwin (1872) drew attention to the fact that the form of submission displays is often diametrically opposite to that of the threat displays of the same species (the principle of antithesis: see section 1.2.2). In the black-headed gull (*Larus ridibundus*), the brown face-mask is directed towards a rival during threat and away during submission or appeasement. Common features of submissive displays are that they involve the concealment of weapons and of anatomical features accentuated during threat and the exposure of vulnerable parts of the body, such as the throat and belly. Morton (1977) has applied a similar argument to avian and mammalian calls (see section 1.2.2).

Pied wagtails (*Motacilla alba*) defend winter feeding territories (Davies 1981). When an intruder lands on a territory it calls 'chis-ick' and the territory holder replies 'chee-wee'. The contexts in which these two calls are used suggest that 'chee-wee' is a signal denoting territorial status and that 'chis-ick' is an essentially submissive signal that an intruder uses to find out if a territory is occupied. In experiments using tape-recorded calls, Davies found that territory owners reacted more aggressively to 'chee-wee' than to 'chis-ick' calls played on their territories.

2.8.3   Signalling sexual motivation

Animals typically perform a variety of displays in association with mating; these are referred to collectively as *courtship displays.* It is important to realise that courtship fulfills a variety of functions, of which signalling motivational state is only one. Tinbergen (1953) summarised the functions of courtship under four headings.

(i) Orientation. Many courtship displays enable individuals to attract mates, often over a considerable distance.

(ii) Persuasion. Typically, males are sexually more active than females when they first meet and male displays may serve to elicit a sexual response from the female. In some instances, the effect of male displays may be to suppress female responses that would hinder successful mating, such as aggression, escape and canni-balism.

(iii) Synchronisation. For many animals it may be important that mating occurs at a particular, optimum moment. This is especially so in species that have external fertilisation, where there is a risk that gametes may be dispersed before they can meet. Sexual signals may serve to coordinate the activities of males and females so that they release their gametes at the same time.

(iv) Reproductive isolation. Courtship displays are often very important in ensuring that animals mate only with conspecifics (see section 2.3).

To these four functions we must add the possibility that courtship displays convey information to potential mates about an individual's quality (see below) and about an individual's sexual motivation.

The communication of sexual motivation can be considered at two levels: long-term changes in behaviour associated with animals coming into reproductive condition over the course of a breeding cycle, and short-term variations associated with the immediate motivation of an individual.

Long-term changes in sexual motivation

Among primates, females typically show regular cycles of sexual activity which, depending on the species, occur throughout the year or for only part of it. For a short time within each cycle, females show an increase in sexual receptivity and, in some species, develop conspicuous signals at this time. For example, chimpanzees (*Pan troglodytes*) have an area of bare skin around the vaginal opening which swells up and turns bright pink, becoming most conspicuous at mid-cycle, around the time that ovulation occurs. Other primates, such as the vervet monkey (*Cercopithecus aethiops*), do not develop such signals. Why some primate species show these anatomical signals, whereas others do not, is a question to which there is no satisfactory answer (Clutton-Brock & Harvey 1977; Chalmers 1979). It is possible that in species that lack sexual swellings, females signal their oestrous condition in other ways, such as by producing characteristic odours or by changes in their behaviour towards males (see section 1.4.3). The production of pheromones that signal oestrous condition is widespread among mammals (Brown 1979).

Among birds, it is very unusual for females to change their appearance with the onset of the breeding season, but quite common for males to do so. For example, the male ruff (*Philo-machus pugnax*) during the winter has plumage similar to the female's but develops elaborate plumes around his head as the breeding season approaches. For many male birds, the onset of the breeding season is a time of greatly increased singing, and, in most species, males do not sing at all at other times of year.

Short-term changes in sexual motivation

For many animals, the mating act leads to a marked reduction in sexual motivation, and there is a refractory period before an individual is motivated sufficiently to mate again. In the smooth newt (*Triturus vulgaris*), males appear to require several hours to replenish their supply of spermatophores (Halliday 1976). Thus, when a male meets a female, he may be at any point in his short-term cycle of sperm depletion and replenishment. On meeting a female, a male newt displays to her, using three distinct tail movements called wave, whip and fan. The rate at which he performs these three displays is strongly correlated with how much sperm he has available, as assessed by the number of spermatophores he produces following a period of display (Fig. 2.8). Thus, the more sperm a male has available, the more vigorously he displays to a female.

In the guppy (*Poecilia reticulatus*), a male's display behaviour is influenced, not only by his sexual motivation, but also by the

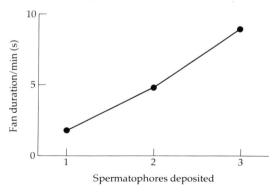

Fig. 2.8. Relationship between the intensity of a male smooth newt's (*Triturus vulgaris*) fanning display and the number of spermatophores that he subsequently deposits. (From Halliday & Houston 1978.)

stimuli that he receives from the female (Baerends *et al.* 1955). Males are more likely to perform their various courtship displays at certain times in their reproductive cycle than at others, and are more likely to display to larger females, which carry many eggs, than to smaller, less fecund females. Thus, the vigour of a male guppy's display is a measure both of his internal state and of his perception of the female as a 'desirable' mate.

Signalling sexual acceptance or rejection

As a consequence of the displays that a courting male directs towards her, a female will either accept or reject him as a mate, thus signalling her own sexual motivation. In many animals acceptance or rejection may be signalled simply by passivity, but in others specific signals are involved. For example, a female black grouse (*Lyrurus tetrix*) crouches in front of that male within a lek whom she has chosen (Kruijt & Hogan 1967), and in many fireflies receptive females reply to the flash pattern of a conspecific male with a characteristic flash pattern of their own (Lloyd 1966, 1971).

The female of the ground beetle *Pterostichus lucublandus* rejects males in a particularly violent way, spraying them with a toxic liquid usually employed against predators, which puts the male into a coma for several hours (Kirk & Dupraz 1972). Female butterflies of the genus *Collias* become unreceptive to males during egg-laying and, if courted at this time, perform ascending flights, sometimes to a height of 20 m, which deter males and enable them to resume oviposition in peace (Rutowski 1978).

In many animals the relationship between a male and a female continues long after they have accepted one another and mating has occurred. This is especially true of monogamous species that form long-lasting pair bonds. Exchange of signals is often involved in the maintenance of pair bonds and mates may not only learn the characteristics of their partner's signals, but may also modify their own to match or complement those of the mate (see section 2.5).

Courtship displays and mate quality

In recent years considerable attention has been given to the idea that animals should not mate indiscriminately with any member of their species, but should choose their mates according to some criterion of mate quality (Bateson 1983). A female, for example, presented with a number of available males, should choose the one holding the most resources, the one who will be the most attentive parent, or the one whose genotype will best complement her own. The nature of the criterion will vary between species and will depend on the nature of the contribution that males make towards reproduction (Trivers 1972; Halliday 1978, 1983). Trivers (1972) has suggested that courtship displays, in addition to the functions listed above, may provide a means by which animals can assess mate quality. As we saw in the case of newts and guppies, the vigour of a male's display is correlated with his sexual motivation which, at least in newts, is a function of his fertility (sperm supply). However, at present there is rather limited evidence for a correlation between male display performance and any other measure of male quality (Halliday 1983). Male mocking birds (*Mimus polyglottos*) with larger vocal repertoires have larger and better quality territories (Howard 1974), and the amount of prebreeding song produced by male stonechats (*Saxicola torquata*) is correlated with the share, relative to that of the female, that they take subsequently in parental care (Greig-Smith 1982).

False information in courtship

Sex is typically a competitive activity, because receptive members of one sex, usually females, tend to be scarcer than sexually active members of the other sex. In some species, competition between males for access to females may take the form of female mimicry, in which males adopt physical features or behaviour patterns characteristic of females in order to mislead rival males.

Small male bluegill sunfish (*Lepomis macrochirus*) are unable to compete with larger males for nesting territories (Dominey 1981). Instead, they adopt the barred colour pattern characteristic of sexually active females and, behaving like females, are able to approach the nests of males engaged in mating with real females. They then swim close to a spawning pair and shed their sperm on to the eggs of the real female. As a result, a small male gains fertilisations at the expense of a territorial male.

During courtship in the scorpionfly *Hylobittacus apicalis*, a male will only be accepted by a female if he presents her with a dead arthropod which she eats during copulation (Thornhill 1979).

Normally, males catch these nuptial gifts, but sometimes males without prey adopt female behaviour patterns and dupe a prey-carrying male into giving up his prey. The mimic is then able to use the prey to attract a real female. Males that steal prey in this way mate more frequently than normal males and do not incur the costs, in terms of effort and risk of predation, involved in prey search and capture (see section 5.4.1).

In many salamandrid amphibians, the female signals her responsiveness to a male by touching his tail with her snout. He responds by depositing a spermatophore over which she walks (Halliday 1977; Arnold 1977). A male who does not have a female to court may show *sexual interference*, moving between a courting pair at the crucial moment and touching the male's tail so as to elicit spermatophore deposition. Having mimicked the female's behaviour he is then able to inseminate the female himself, either by depositing his own spermatophore on top of the first male's, or by leading her away and initiating his own courtship (Arnold 1976).

In all these examples, signals normally performed by females are adopted by males to gain an advantage over their rivals.

2.8.4 *Motivational conflict*

There are numerous occasions in an animal's life when it will be motivated to perform more than one activity. If stimuli relevant to two incompatible activities are strong at the same time, an animal may perform neither activity but shows some other form of behaviour that is interpreted as being the expression of its motivational conflict. Its behaviour may be ambivalent between the two conflicting patterns, or it may perform some apparently irrelevant action, called a displacement activity (Tinbergen 1952b; Baerends 1975). Particular forms of conflict behaviour are often characteristic of specific motivational states; they thus have considerable potential as signals conveying information about the motivational state of an animal.

Behaviour whose causation can be interpreted in terms of motivational conflict is especially common during social interactions between animals because incompatible tendencies, such as approach and withdrawal or sex and aggression, are often aroused simultaneously by the presence of the social partner. For example, during courtship in the chaffinch (*Fringilla coelebs*) a male adopts a distinctive head-down posture in which he stands broadside to the female. This appears to be an expression of ambivalence between tendencies to approach and to flee from the female (Hinde 1953).

The potential communicative value of conflict behaviour is recognised in theories about ritualisation (see section 2.2). Many displays are believed to have evolved from forms of conflict behaviour, and to have become more conspicuous, elaborate and stereotyped in the process.

2.9   Communication about the environment

So far this chapter has discussed communication between animals that conveys information about some aspect of the performer. In many species, signals are used to transmit information about various aspects of the external environment.

Animals may discover a lot about their environment simply by observing one another, without exchanging specific signals. For example, a bird feeding busily on a lawn may attract other birds who, from the behaviour of the first bird, are alerted to the presence of a rich food supply. A number of authors (Ward & Zahavi 1973; Krebs 1974) have suggested that a colonial nesting site may act as an 'information exchange' for birds which, by observing the directions in which successfully foraging individuals fly into and out of the colony, may discover the location of food sources. Although such examples involve the passing of information between individuals, they cannot be called communication in the proper sense of the word, since they do not involve signals the function of which is to convey information (see Chapter 1).

2.9.1   *Alarm calls*

In many passerine birds, individuals which detect a predator, such as a hawk, produce a distinctive alarm call (Marler 1955; section 1.2.2). This call is very similar across several species and consists of a short-duration (*c.* 0.5 s), high-frequency (7 kHz) note. A sound with these characteristics is very difficult to locate and this appears to be the selection pressure that has led to convergent evolution in several species. When they hear an alarm call, other birds fly into the nearest cover.

Some birds have more than one alarm call, each used in response to a different kind of predator. For example, many birds use an alarm call of the kind just described in some contexts but in others use an easily located mobbing call that attracts other birds who together attack the predator. The vervet monkey (*Cerco-pithecus aethiops*) has several distinct alarm calls, each signalling the presence of a different predator (Seyfarth *et al.* 1980; section 1.2.1). Animals hearing these calls make appropriate responses, for example running into trees in response to a leopard call, looking up in response to an eagle call and looking down if they hear a snake call.

2.9.2   Bee language

The classic example of communication of information about the outside world is the language used by honeybees to convey to fellow hive-members the exact location of food (von Frisch 1967;

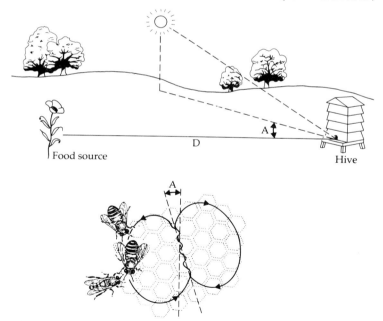

Fig. 2.9. The waggle dance of the honeybee (*Apis mellifera*). Top: the relationship between hive, food source and sun. Below: a worker performing the dance on a vertical comb, with two attendants.

Gould 1976). Bees have two dances, the round dance and the waggle dance. The round dance, in which a worker who has found food simply rushes around in circles, is used when the food she has found is less than about 50 m from the hive. Other workers, using olfactory cues from food that she has brought back with her, simply fly out and search for the food source. For food sources at greater distances, the more elaborate and precise waggle dance is used (Fig. 2.9). This dance is described as a *symbolic language,* because the dancer encodes specific kinds of information

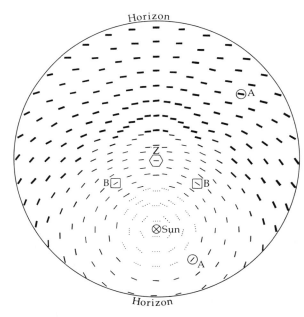

Fig. 2.10. Schematic representation of a bee's view of the sky when the sun is at an elevation of 45°. The dark bars show the angle and degree of polarisation of light. The thickness of the bars is proportional to the degree of polarisation; the dotted bars near the sun indicate that polarisation is below the perceptual threshold of bees. The angle of polarisation is plotted with respect to the horizon. Points such as those labelled A have similar polarisation angles and elevation, but are different distances from the sun. When shown such a pattern, bees interpret it as being the one further from the sun. Points such as those labelled B have similar angles, elevation and are equidistant from the sun, but are on different sides to it. When shown such a pattern, bees interpret it as being the one to the right of the sun. The point Z is the sky's zenith (the point directly overhead). When shown the pattern at Z, bees are unable to use either rule. (Modified from Brines & Gould 1979.)

in particular aspects of her dance. The dance is performed on the vertical plane of the comb inside the hive, and the angle between the central axis of the dancer's movement and the vertical plane is the same angle (A in Fig. 2.9) as that between her path from the food source to the hive and the sun. In the straight part of the dance, the dancer waggles her abdomen and produces bursts of sound; the number of waggles and of sound bursts is proportional to the distance (D in Fig. 2.9) between the hive and the food source.

In tropical bees that live in open swarms, rather than enclosed hives, and in temperate-zone bees dancing outside the entrance to the hive, the dance is performed in the horizontal plane, and the rule relating the vertical axis to the sun's direction becomes irrelevant. Instead, dancers orient directly to cues in the sky such as the sun, and dance attenders use the same points of reference (Brines & Gould 1979).

Recent studies have shown that the bee dance language is even more sophisticated and complex than so far described. If the sun is visible in the sky, bees have no problem; it provides a single reference point that both dancers and attenders can use. However, problems arise when the sun is obscured by cloud. At such times bees refer to patches of blue sky and, from the pattern of polarised light in such patches, deduce the sun's position by extrapolation. This introduces an ambiguity as any given pattern of polarisation generally exists in two parts of the sky because light is polarised in a symmetrical fashion on each side of the sun (Brines & Gould 1979). Ambiguity exists in two respects; two patches with similar polarisation patterns may be at different distances from the sun and may be on opposite sides of the sun. Thus, when a bee examines a particular patch of sky it does not know which of two possible distances it is away from the sun and whether it is to the left or the right of the sun (Fig. 2.10).

Brines and Gould investigated the way that bees solve these problems by using an observation hive turned so that bees danced on a horizontal plane and illuminated by an artifical light as the only cue available as a dance reference point. The artificial light could be altered so that bees interpreted it as a patch of sky rather than the sun. (Light direct from the sun contains much less ultraviolet light than that from blue sky.) These experiments revealed that bees resolve the ambiguity by using arbitrary rules. Bees dancing in reference to an artificial patch of sky which, on the basis of its polarised light characteristics, could be at one of two distances from the sun, always interpret it as being at the further distance. Presented with a patch that is one of two that could be equidistant from but on different sides of the sun, dancers always interpret it as being to the right of the sun.

Because bees use these arbitrary rules to interpret what they see, they will incorrectly identify a patch of sky half of the time. However, because dancers and dance attendants use the same rules, and because direction to the food source is referenced in the dance to the position of the sun (extrapolated from the patch of sky), not of the patch of sky itself, mistakes are cancelled out and bees recruited by dancers will fly out in the correct direction.

Interestingly, there is one situation where the rules fail. If a patch of sky is on the imaginary line passing through both the sun and the zenith, it may be one of two that are equidistant from the sun, but neither to its left or right. Presented with such a patch, bees dance according to both the possible interpretations in a roughly 50:50 proportion and bees recruited by the dance fly off in the wrong direction in about 50% of cases.

2.10   Interspecific communication

Reflecting the primary concern of ethologists, this chapter has so far dealt entirely with communication between individuals belonging to the same species. However, there are occasions when members of one species communicate with members of another. It is probable that a great deal of information passes between animals belonging to different species but, in the great majority of instances, this is likely to be 'incidental' and does not involve signals that have evolved for that purpose. To illustrate this distinction, let us return to the hawk alarm call of passerine birds (see section 2.9.1).

When an individual great tit gives an alarm call on seeing a hawk, we naturally assume that the information conveyed is 'intended' for other great tits. However, individuals of other species also respond to the call and thus benefit from the information that it contains. The fact that alarm calls confer a benefit on birds other than the caller raises interesting and important questions about their evolution, which can be discussed only briefly here. Maynard Smith (1965) pointed out that, as an example of apparently altruistic behaviour, alarm calls pose problems for the theory of natural selection, in which the adaptiveness of a character depends on its contribution to the fitness of individuals showing that character. He suggested that they may have evolved by kin selection, on the assumption that some of the respondents are likely to be close relatives of the caller. Charnov and Krebs (1975) have suggested that alarm calls may be essentially exploitative, rather than altruistic, and could therefore have evolved by selection at the level of the individual. They argue that a caller passes on only part of the information that it possesses; it signals that a hawk is nearby, but not *where* the hawk is. Its alarm call creates a situation, where many birds are dashing for cover, which is confusing to a predator and in which the caller can make its own escape in the optimal direction. This ploy will be all the more effective if birds of many species respond to the alarm call.

However, although alarm calls may provide an example of interspecific communication, we need not argue that they evolved to fulfil that function. The similarity of the hawk alarm call in several species can be attributed to convergent evolution, because of the properties that make it difficult to locate.

A further possibility is that alarm calls are signals from prey to predator, indicating that the predator has been detected and perhaps causing it to search elsewhere for less alert prey. A similar suggestion has been made for the function of the conspicuous 'stotting' displays shown by various mammals when pursued by a predator (see Harvey & Greenwood 1978).

2.10.1 *Interspecific aggression*

Anyone who has observed birds squabbling at a bird table on a winter's day will be aware that different species seem able to recognise and respond to one another's aggressive displays. This could be the result of each individual learning, through painful experience, what heterospecific threat displays mean, but it is more likely to be due to the fact that threat displays are rather similar across species. As described in section 2.8.1, they tend to involve the prominent display of weapons, an increase in apparent size, and the production of harsh, low-frequency sounds. Again, it seems likely that convergent evolution has led to conditions in which different species can 'understand' one another's signals.

In many territorial species, individuals defend their territories only against conspecifics, but in some they are aggressive to heterospecifics as well and are said to show interspecific territoriality. It has been suggested that such behaviour results from a mis-identification of species and is thus not adaptive (Murray 1971). This view is widely disputed, for example by Kohda (1981) who has shown that in the pomacentrid fish *Eupomacentrus altus* interspecific aggression is directed selectively towards living and model fish whose body shape generally corresponds to that of species which, like itself, are herbivorous and are, therefore, in direct competition for food. Interspecific territoriality occurs between chaffinches (*Fringilla coelebs*) and great tits (*Parus major*) on Scottish islands, where food supply is limited, but not on the mainland, where it is more abundant (Reed 1982). Mainland birds of both species respond only to conspecific song whereas island birds show a territorial response to the song of either species.

2.10.2   Cooperative interspecific communication

For a hermit crab, it is important to live in a shell that fits snugly around the body. Thus, as it grows, it must exchange its present shell for a new, larger one. One way that a crab may acquire a better-fitting shell is by taking over one that is occupied by another crab. This involves quite complex displays in which two crabs perform rocking movements and tap one another's shells with their claws. The end result of these interactions is that the two crabs swap shells. Because the initiator of these exchanges typically gains by acquiring a better-fitting shell, they have been categorised as aggressive behaviour. However, Hazlett (1978) has shown that, in many interactions, both crabs acquire a better shell and he suggests that they should be seen as mutually beneficial 'negotiations', not as aggression. Shell exchanges can occur among crabs belonging to three different species, individuals using the same behaviour patterns that they use in intraspecific exchanges (Hazlett 1983).

Interspecific communication is perhaps most highly developed in symbiotic relationships, such as that between the cleaner fish *Labroides dimidiatus* and its various 'customer' species (Wickler 1968). The cleaner removes bacteria, ectoparasites and diseased or damaged tissue from its client; as a result, one partner derives food and the other is kept healthy. Many customer fish are large predators that pose a potential threat to the cleaner's life, but the symbiosis is maintained by behaviour patterns that enable partners to identify one another correctly. A cleaner signals its preparedness to clean by performing a distinctive 'dance' in which it swims slowly forwards, repeatedly raising and dropping its tail. A customer signals its readiness to be cleaned by adopting a trance-like posture. If a customer is too active for a cleaner to scour it thoroughly, the cleaner jabs it with its open mouth. Some individuals of territorial customer species are cleaned by particular cleaner individuals and Wickler suggests that such relationships may involve individual recognition.

2.10.3 *Exploitative interspecific communication*

Just as intraspecific patterns of communication may be exploited by animals that use signals in 'false' contexts (see section 2.8.3), so interspecific signals may be used by predators and parasites. A predatory fish, *Aspidontus taeniatus,* is a remarkably good mimic of the cleaner fish described above, not only bearing a very close physical resemblance, but also mimicking the cleaner's distinctive 'dance'. Having deceived and got close to a customer fish, it takes a bite out of it (Wickler 1968). It appears, however, that as they get older, customer fish are able to learn the difference between cleaners and their predatory mimics.

The sexual signals of fireflies of the genus *Photinus* are exploited by predatory females of another genus, *Photuris*; these *femmes fatales* mimic the sexual signals of their prey to lure male *Photinus* to their death (see section 3.6.2).

Examples of interspecific mimicry are not confined to visual signals. A variety of arthropods, especially beetles, are able to inhabit ant nests and to be fed by their hosts by mimicking ant odours and tactile signals (Holldobler 1971).

2.11   Measuring information

So far in this chapter, the word 'information' has been used in its rather vague everyday sense. It can, however, be used in a more precise way that enables ethologists to analyse communication between animals quantitatively. Defined formally, information is said to have passed from a performer to a receiver when the performer's behaviour becomes more predictable to the receiver. In the practical application of information theory, predictability is expressed inversely as a reduction in uncertainty. To quantify information we require standard units, and information is measured in *bits*. Suppose that we find an animal and are uncertain about its gender. We can reduce our uncertainty by asking just one question, 'Is it female?', and looking for the diagnostic features of a female. Since the answer to this question is 'yes' or 'no', only one question is required to eliminate our uncertainty and, by definition, one bit of information has been passed. A question for which there are four possible answers can be solved by passing two bits of information, one with sixteen possibilities requires four bits, and so on.

In the quantitative analysis of communication between two animals we must introduce an additional component, a non-participant observer, in this context an ethologist (see section 5.2.1). When two animals, A and B, are observed, the less un-certain the observer is about what B will do following specific actions by A, the greater is the reduction in the observer's uncertainty and, therefore, the greater is the amount of infor-mation transmitted to the observer (Shannon & Weaver 1949; Attneave 1959). Uncertainty (*H*) is defined as follows:

$$H = - \sum_i p_i log_2 p_i$$

where there are *i* possible patterns of behaviour and p_i is the probability that the *i*th of them will occur. The use of logarithms to the base 2 arises from the convention that the reduction of un-certainty involves binary questions, to which there are only two possible answers, yes and no.

If animals A and B are not simply behaving independently, the value of *H* for B's behaviour should decrease, the more we know about A's. In other words, when A performs a particular action, the observer should be able to predict B's subsequent behaviour reasonably reliably. Expressed formally, the reduction in uncer-tainty perceived by the observer (H_T), is given by

$$H_T = H_R - H_{R|S}$$

where H_R is the observer's uncertainty about the receiver's (B's) behaviour when the signaller's (A's) behaviour is not known and

$H_{R|S}$ is the uncertainty perceived when the signaller's behaviour is known.

The value of H_T can be derived from detailed analysis of the sequences of behaviour performed by A and B. This has been done for encounters between hermit crabs (Hazlett & Bossert 1965) and between stomatopod crustaceans (Dingle 1972). We must be very cautious in any interpretation we put on the numerical value of H_T. Information theory can be very useful in determining whether or not the behaviour of one animal is influencing that of another. If H_T is large, it follows that H_R is much greater than $H_{R|S}$ and, therefore, that the receiver's behaviour is strongly dependent on the signaller's. Conversely, if two animals are behaving independently of each other, H_T will be small. As long as the value of H_T is greater than zero, we can be sure that the behaviour of two animals is correlated in some way; this correlation may or may not involve communication in the formal sense defined in Chapter 1.

However, problems arise if we try to read too much into the precise value of H_T, because its value will depend on the value of i, the number of possible patterns of behaviour that we observe. If the animals under consideration differentiate more distinct patterns of behaviour than we do as observers, our estimate of i will be too small and our value of H_T will correspondingly be an underestimate. This brings us back to a point stressed earlier (section 2.2); our attempts, as observers of behaviour, to categorise signals may bear little relationship to the variety of signals perceived and discriminated between by animals. The accuracy or otherwise of estimates of H_T will be especially important if comparisons are being made between species, because i may be estimated more accurately in one species than in another. For similar reasons, studies of the same species by different researchers may be difficult to compare.

It must be emphasised that this quantitative approach to communication measures the information that is transmitted from interacting animals to a human observer. It estimates the extent to which the behaviour of one animal is dependent on another's, and may thus be a consequence of information that has passed between them. It is not, however, a direct measure of that information, and, as mentioned above, the dependence is statistical and need not necessarily imply that one animal has influenced the other in any way (see also section 1.3.2).

2.12   Conclusion

In a chapter such as this, it is possible to give only a glimpse of the rich variety and complexity of animal communication and to mention only a few of the numerous studies that have been carried out. The fact that this chapter has treated animal communication under a series of discrete functional headings may have created the false impression that each animal signal fulfils a specific function. For many signals, this is far from true. A given signal may convey more than one kind of information to a single receiver, or different kinds of information to different receivers. For example, the characteristic calls of male frogs used to be known as mating calls because they attract females but, with the realisation that they serve as aggressive signals between males, they are now referred to by the less specific term of advertisement calls. Ethologists have a natural tendency to classify the phenomena they observe into neat categories. However, to label the scent mark of a mongoose simply as a threat display does scant justice to the richness of its communicative power. Not only does it signal that the marker is aggressive, but also how aggressive it is, which individual made the mark, and when it did so. Identifying the functions of a particular signal is no easy task; for each putative function, a specially designed set of carefully controlled experiments, which pay particular regard to the contexts in which signals are performed, must be conducted (see Chapter 1). As more and more such studies are carried out, we may expect to discover that animals are capable of passing among themselves information of greater variety and complexity than previously we had imagined.

2.13   Selected reading

An excellent introduction to the subject of communication is provided by Cullen (1972). This is a chapter in a book, *Non-verbal Communication*, edited by R. A. Hinde, which contains a number of useful contributions that discuss communication in a variety of animals; there are also several chapters on human communication. *How Animals Communicate*, edited by Sebeok (1977), contains an immense wealth of information and is essentially a work of reference. W. J. Smith's book, *The Behavior of Communicating*

(1977), discusses signals according to the messages they convey, but does so from a somewhat individual standpoint. A useful introduction to the principles of information theory is Attneave's *Applications of Information Theory to Psychology* (1959).

CHAPTER 3
COMMUNICATION
AND THE
ENVIRONMENT

H. CARL GERHARDT

3.1 Introduction

Every animal signal is a form of energy: chemical, electromagnetic, mechanical or electrical. The form of a signal is thus constrained by physical laws governing its production and effective transmission through natural environments. Biological limitations are imposed by anatomy and by the fact that the energetic costs of signalling can be high. Furthermore, different kinds of animals often signal in the same place at the same time. Competitors may interfere with communication and predators may exploit it, turning signallers or responders into prey. This chapter deals with the physical and biological factors that affect intraspecific communication between spatially separated individuals.

For communication to occur, signals usually must be detected and recognised. Detection involves a decision about whether or not a signal has occurred. Recognition implies that a detected signal has been categorised, and this is usually indicated by the particular response that the signal elicits. If a sensory system is very narrowly tuned, so that only signals with very specific properties excite it, then detection and recognition may be almost synonymous. More often an animal must process a number of potential signals before recognition occurs and, especially if the noise level is high, it may have to move closer to their source to do so. Even if localisation is not a prerequisite for recognition (or occurs nearly simultaneously, as in unobstructed vision), most responses are oriented with respect to the origin of the signal. In this chapter we will consider how environmental factors limit the effective detection, recognition and localisation of animal signals. Stated another way, it will be shown how animals produce signals which facilitate these processes in particular environments.

3.2 Chemical communication

Chemical communication is nearly ubiquitous in animals, and it was almost certainly the first kind of communication to evolve. We shall focus here on signalling by pheromones, a term derived from the Greek words *pherein* (to carry) and *hormon* (to excite). Pheromones may be bioenergetically cheap signals. Wilson and Bossert (1963) theorised that most chemical signals should have 5 to 20 carbon atoms and molecular weights of about 80 to 200. They reasoned that some minimum complexity is necessary for specificity and effective stimulation of the receiver's receptors; the upper limit is probably set by a reduction in volatility and by the energetic costs of biosynthesis and transport. Many pheromones have now been characterised, and these predictions are generally confirmed, at least for airborne signals. Chemicals used for communication in water, however, may be very much smaller (hydroxyl ions are possibly used as pheromones in nematodes) or larger (proteins are used by some snails) than airborne pheromones (Wilson 1970; Shorey 1976). Since specificity can also be accomplished by mixtures of chemicals, individual compounds can be rather simple in structure (see below).

3.2.1 *Communication distance*

Animals usually release pheromones into the environment by opening or everting a storage gland or secretory skin surface. Urine and faeces are also used by many kinds of mammals to carry a pheromone to the exterior, where it marks a substrate and subsequently evaporates.

When the medium into which a pheromone is emitted is stationary, dispersion occurs by diffusion. This is a very ineffective mechanism. Shorey (1976) estimates that only about 10% of the molecules of a typical airborne pheromone (molecular weight of 200) travel 1 cm from the source in 1 second. Diffusion in water is 10^4–10^5 times slower than in air. As pheromone molecules spread farther from the source, their concentration soon becomes so small that potential receivers no longer detect or respond to the signal. Bossert and Wilson (1963) deal with pheromone dispersion in time and space by developing equations to estimate theoretical active spaces. An active space is the area containing an above-threshold concentration of pheromone; its volume depends on the number of molecules emitted (Q), the receiver's threshold sensitivity (K), the mode of emission (discrete puff or continuous), pheromone diffusion rate, and movement (if any) of the medium. The limitations of diffusion can be appreciated by considering the broadcast of an alarm pheromone by the harvester ant, *Pogomyrmex badius*. If emission occurs as a single puff into still air, the maximum radius of the active space is about 6 cm; continuous emission increases the effective distance to about 1 m (Wilson & Bossert 1963). Even very large Q/K ratios (large number of molecules emitted and very sensitive recipients) result in active spaces that extend slowly for limited distances (Table 3.1).

The dynamic properties of an active space are also very important. The time required for the active space to reach its maximum volume is mainly dependent on pheromone diffusion rate and the Q/K ratio; just as significant is the fade-out time, or time after emission when the active space no longer exists. Bossert and Wilson (1963) show that lowering Q, raising K or increasing the volatility of the pheromone reduces the fade-out time. They argue further that short fade-out times, although limiting transmission distance, provide more information about the time of release and allow the source to be more easily localised because the concentration gradient of the pheromone will be fairly steep. A puff of harvester ant's pheromone, for example, reaches its maximum range in about 13 seconds and fades out completely about 22 seconds later. This is a desirable property for an alarm signal, since it does not persist after a disturbance is over and so the social group

Table 3.1. The predicted spread and fade-out (active space) of a signal comprising a substance of low molecular weight*. (From Wilson 1970.)

$\dfrac{Q}{K}$	Diffusion coefficient		Maximum radius of active space (cm)		Time required to reach maximum radius (s)		Fade-out time (s)	
	Air	Water	Air	Water	Air	Water	Air	Water
1	10^{-1}	10^{-5}	0.6	0.6	0.4	4×10^{3}	1	10^{4}
10^{2}	10^{-1}	10^{-5}	2	2	8	8×10^{4}	20	2×10^{5}
10^{4}	10^{-1}	10^{-5}	10	10	1.5×10^{2}	1.5×10^{5}	5×10^{2}	5×10^{6}
10^{6}	10^{-1}	10^{-5}	60	60	4×10^{4}	4×10^{7}	10^{4}	10^{8}

*Diffusion coefficient arbitrarily set at 0.1 cm s^{-1}.

soon resumes normal activity. The dynamic properties of the active space of a hypothetical pheromone as a function of the Q/K ratio are shown in Table 3.1.

How then can an animal overcome the limitations of diffusion? If molecules move only slowly and for limited distances from a stationary source, one solution is to broadcast pheromones while moving. Depending on fade-out time, an animal can lay down a durable trail that can easily be followed. Many insects do just this. The fire ant, *Solenopsis saevissima*, makes short-lived (less than two minutes) trails under a metre long to mark transient food sources. Recruits renew the trail as long as the food source persists. After the food is removed, the trails quickly disappear and the ants forage elsewhere. By contrast, Texas leaf-cutting ants have long-lasting (several months) trails to permanent sources of vegetation more than 100 m from their nests; both volatile and non-volatile pheromones as well as visual cues (the cleared trail) are used to guide them (Blum 1974).

Another, more common, solution to the diffusion problem is to broadcast pheromone into a moving medium, a wind or current. If the velocity of the medium is too high, however, there is little gain. Turbulence disperses molecules laterally relative to the direction of flow and so decreases the volume of the active space downstream. Furthermore, flying or swimming recipients may not be able to make sufficient progress toward the source if the wind or current is too strong. Once again, Bossert and Wilson (1963) developed a theory (from micrometerology) and calculated theoretical active spaces as a function of wind velocity for the airborne pheromone (bombykol) of the gypsy moth. As Farkas and Shorey (1974) pointed out, however, real pheromone plumes hardly resemble the calculated semi-ellipsoid spaces depicted by Bossert and Wilson; rather, they must be something like the swirling, filamentous trails we can see when smoke is wafted from a burning cigarette (Fig. 3.1a). Estimates of maximum theoretical distance (about 100 m) for pheromone communication in the cabbage looper moth (*Trichoplusia ni*) are lower than those calculated by application of the Bossert–Wilson formulae when other factors such as duration of pheromone emission and approach speed of the male recipient are considered (Sower *et al.* 1973). Furthermore, Bossert and Wilson may have greatly overestimated the sensitivity of the male gypsy moth and hence the maximum theoretical range (*c.* 4 km) of the female pheromone under favourable wind conditions; results of release and recapture experiments to verify the predictions are not yet conclusive (Shorey 1976).

Kaae and Shorey (1972) discovered that female cabbage looper moths appear to adjust the duration of their pheromone broadcast according to prevailing wind velocity. Broadcast time is increased

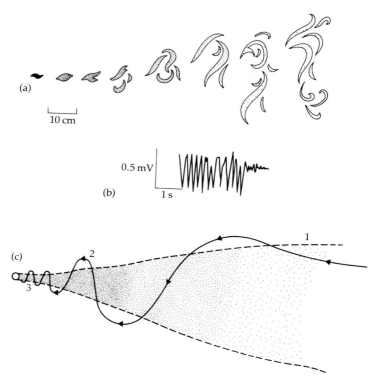

Fig. 3.1. (a) Artist's conception of a temporally patterned pheromone plume. (Based on Conner *et al.* 1980 and Farkas & Shorey 1974.) (b) Electrical response of the antennal chemosensory organ of a male moth (*Utetheisa ornatrix*). The preparation was situated about 6 cm downwind from a scenting female, and the rhythmic changes in neural activity reflect the temporal pattern of the female's output of pheromone. (c) Hypothetical flight path of an insect locating a pheromone source by (1) first orienting upwind in an area of low concentration and wide active space; (2) flying out of the plume and losing the trail and then using crosswind flight to relocate the trail; (3) entering an area of high concentration where visual orientation is activated and flight tends to be arrested. (Adapted from Farkas & Shorey 1974.)

from about 5 minutes at 3 ms^{-1} to about 20 minutes when wind speed falls below 0.1 ms^{-1}. The increase extends the maximum range at low wind velocity; the upper limit is probably set by the ability of the male to fly upwind at reasonable speeds.

3.2.2   Source localisation

Some authors suggest that pheromone gradients arising solely from diffusion may be steep enough for source location at close range (Bossert & Wilson 1963; Farkas & Shorey 1972, 1974; but see Kennedy & Marsh 1974). The searching animal would presumably sense the pheromone gradient chemotactically by successive comparisons of concentration in different points in space, or (even less likely) by instantaneous comparisons of the responses of paired receptors (e.g. antennae in insects). A moving medium simplifies the problem, and it is well established that insects locate pheromone-emitting partners by orienting to the wind direction and moving upwind (anemotaxis). Aquatic organisms presumably use currents in the same way (rheotaxis). Since the trail is likely to be discontinuous, the animal must periodically lose contact with the pheromone. One strategy, observed in moths, is then to make back and forth movements crosswind, until the trail is again detected and upwind orientation resumes (Fig. 3.1c). Insects also use visual optomotor reactions to the apparent movement of the ground to orient themselves in the wind (Kennedy & Marsh 1974). Tobin (1981) has recently shown that although the zig-zag path of cockroaches is regulated by the width of a pheromone plume, the insects make many turns before reaching its edge. He suggests that there is an additional, internal control mechanism for turning that is initiated but not directed by exposure to the pheromone.

Conner *et al.* (1980) established that females of the arctiid moth *Uretheisa ornatrix* pulse the emission of the principal component of their pheromone mixture at about 1.5 puffs s^{-1} (Fig. 3.1b). This is the first example of temporal patterning in pheromone communication, despite Bossert's (1968) theoretical demonstration of the potential advantages in terms of localisation and information transfer. Pulsing a signal is equivalent to reducing its fade-out time but without the disadvantage of having to reduce threshold sensitivity (one way of lowering the Q/K ratio). Conner and his colleagues estimate that the temporal pattern remains intact over at least 60 cm; they suggest that a male detecting the pulsing could then change his search pattern from upwind orientation to some close-range mechanism, such as vision. Discontinuous exposure to the pheromone also may help to avoid prolonged adaptation of the male's chemoreceptors and hence habituation. Such receptors lose their sensitivity if exposed for long periods of time to a chemical stimulus. Indeed, the zig-zag (crosswind) flight patterns of other insects may not be merely orientation manoeuvres but may also be a mechanism to maintain chemosensitivity. Wilson and Bossert (1963) also suggested this second benefit in connection with the sinusoidal movements of insects following terrestrial trails.

3.2.3   *Recognition of chemical signals*

The most important form of environmental noise that a chemical communication system must overcome is the presence of phero- mones of other species. The chemosensory systems of many male moths are so finely tuned that even slight differences in chemical structure (isometric changes, for example) drastically reduce or eliminate perception. Other foreign pheromones may be detected but cause inhibition or even repulsion of the recipient.

So far discussion has been simplified by considering one kind of pheromone per species. In reality, insects and other animals almost always produce blends of chemical signals, and it is likely that the female moth produces a spatially patterned plume which drifts downwind and offers sequential cues to a male moving towards her. Chemical medleys almost certainly decrease the chances of attracting an individual of another species. Even if closely related species use the same pheromones, the ratio of components, and even absolute concentrations, are usually species-specific (Roelofs 1979; section 2.3).

Although chemical specificity is well documented, cross- species attraction by pheromones is also widespread (Ganyard & Brady 1972; Nielsen & Balderston 1973; Shorey 1976). Mismating is avoided by differences in geographic distribution, in habitat, or in the timing (seasonal or time of day) of courtship. The alfalfa looper moth (*Tricoplusia ni*) and the cabbage looper moth (*Autographa californica*), for example, appear to release the same sex phero- mone, but they tend to do so at different times of night (Shorey

1976). If these mechanisms fail, then differences in short-range courtship behaviour or morphological or genetic incompatibility prevent hybridisation.

3.3   Visual communication

Most visual communication depends on ambient light, ultimately derived from the sun. The environment modifies the amount and quality of sunlight primarily by absorbing and scattering some wavelengths more than others. Passage of light across an air–water interface is governed by laws of geometric optics; reflection and refraction (bending) at such boundaries are important for organisms living in both media and for image formation by many kinds of eyes.

Some insects can see ultraviolet wavelengths as short as 280 nm; the longest wavelengths (the red end of the spectrum) that effectively excite visual receptors are around 700 to 800 nm. Both limits appear to be set by the photochemistry of light-absorbing molecules used in vision; radiation of shorter wavelength can also damage biological tissues. Except for the removal of short-wavelength radiation by the ozone layer, the atmosphere narrows the visible spectrum of sunlight only moderately. Filtering of sunlight is much more pronounced in water, which, even if clear, absorbs and scatters long wavelengths very strongly. Light scattering is also important in air. If the particles responsible for scattering are small relative to the wavelengths of light affected, then short wavelengths are more strongly scattered than long ones, a phenomenon known as Rayleigh scatter. The blue colour of a clear sky or a body of water is mainly attributable to such scatter.

3.3.1   The 'quit point'

Vision is obviously limited at night, when even bright moonlight is 10 log units (10^{10} times) less intense than daylight. In some environments, such as caves and deep or turbid water, sunlight is absent or so weak that passive visual communication is impossible or nearly so. A dramatic illustration of this fact is the change in the fauna, which, depending on the light-transmission properties of the ocean, occurs at depths of 700 to 1300 m (Lythgoe 1979). The organisms in the zone just above this 'faunal break' generally have large eyes and are often silvery, transparent or countershaded. Bioluminescence is common. These are characteristics of animals that hunt, avoid predators, and communicate by vision in dim light. At greater depths, organisms have poorly developed, often regressed eyes and dull, non-reflecting body colouration. Other animals that have abandoned vision as a major sensory modality include many bats, cave fish and salamanders, cave invertebrates, and fishes living in very muddy water. These creatures have developed other senses, sometimes exotic ones, to a striking degree: echolocation in bats, displacement (water current) detectors in aquatic cave-dwellers, and electrolocation and electric communication in mormyrid and gymnarchid fishes (see Chapter 4).

3.3.2   Changes in light quantity and quality

Some animals are active both by day and by night, and the change in light intensity from daylight to just after sunset can be as much as a log unit (tenfold) per minute. Other animals, such as diving mammals, move from well-lit environments to dim or even totally dark regions in a matter of minutes. Many visual adaptations have evolved to deal with these changes, including dual pigment systems, neural adaptation, mechanisms to alter the amount of light reaching photoreceptors, and even changeable filters.

As the sun sets, there are also changes in light quality. Moonlight and starlight are richer in long wavelengths than daylight, but there is a reduction in the yellow–orange part of the spectrum around twilight, a phenomenon called the Chappuis effect (Fig. 3.2a). The well-known Purkinje shift, from cone to rod vision as light level decreases, seems to be most advantageous for the many animals active at twilight (Munz & McFarland 1977). The change in maximum visual sensitivity from around 550 nm (cone vision) to about 500 nm (rod vision) would certainly seem to match the spectrum of twilight much better than that of night.

Fig. 3.2. (a) Comparison of spectral irradiance during midday, twilight, and a moonlit night at Einewetok Atoll. (From Lythgoe 1979, after McFarland & Munz.) (b) Downwelling irradiance in Crater Lake, Oregon. The natural water is exceptionally pure and clear. Calculated values are based on measurements at 10 m. Note the progressive filtering as depth increases. (After Lythgoe 1979.)

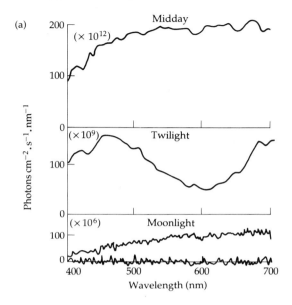

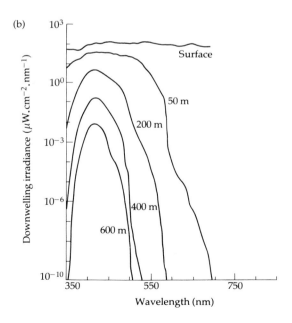

3.3.3   *Light quality in water*

As mentioned above, selective absorption and scatter together determine the amount and quality of light available for vision underwater (Jerlov 1976). Pure water best transmits light of about 475 nm; it filters out shorter and longer wavelengths, especially those at the red end of the spectrum above about 575 nm. This window to short wavelengths becomes narrower as depth increases (Fig. 3.2b). The absence or reduction of long wavelengths in deep water changes the apparent colour of objects. Fish that are bright yellow or red near the surface appear greenish in deep water. The short-wavelength bias of deep, clear water is correlated with the near-universal shift in the spectral sensitivity of the visual photopigments in marine invertebrates, fishes and mammals (Munz & McFarland 1977; Lythgoe 1979). Fishes that live in deep water but still use ambient light for vision, for example, have photopigments with absorption maxima around 480 nm; whereas fishes of shallower, coastal areas are maximally sensitive around 500 nm. Another solution, adopted by migratory fishes such as salmon, is to have two kinds of photopigments, one optimal for deep sea existence and the other for shallow fresh water. The balance in abundance of the two pigments changes appropriately during migration. Indeed, the myriad of different (especially cone) photopigments being discovered as techniques are refined is probably an evolutionary consequence of the advantages of broader spectral sensitivity in many aquatic environments. Lythgoe (1979) suggests that the function of these photopigments may be to enhance contrast perception, i.e. differences in apparent brightness between objects of interest and the background (see below), rather than merely increasing overall visual sensitivity.

3.3.4   *Light quality on land*

Visual communication on land is seriously constrained by light scattering, which creates a visual veil between objects of interest and the observer's eye. This veil reduces contrasts in brightness and colour that are necessary for image formation and pattern recognition. In many situations the shorter wavelengths are scattered more than other wavelengths, which may therefore constitute the bulk of the image-forming light. Diurnal animals and those living in shallow water seem to have adopted the obvious solution: the insertion in front of the photoreceptor surface of a filter which selectively reduces the transmission of shorter wavelengths. Such 'yellow filters' also reduce scatter within the eye and filter out potentially damaging ultraviolet light (Tansley 1965; Muntz 1975). Nocturnal animals generally lack yellow filters, and oil droplets or other shielding devices are poorly developed, if present at all. These animals can hardly afford to lose any of the scarce photons available. Scattering from large particles such as water droplets in fog equally affects nearly all wavelengths and thus poses a problem for visual communication for which there is no good solution.

3.3.5   *Directional distribution of light*

Light from the atmosphere is reflected and bent at the water surface because of the difference in the optical density of these two media. The 180° solid angle that includes the sky is consequently compressed into a 97° solid angle called Snell's window, or a fish-eye's view of the aerial world (Fig. 3.3). This area of light extends to about 30 m in clear water and, in shallow waters, light reflection from the bottom also can be an important source of illumination. In deep water, about 98% of the light comes from

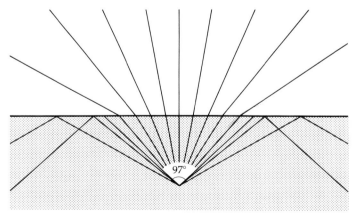

Fig. 3.3. Patterns of light refraction at the surface of the water which give rise to what is termed 'Snell's window'. A fish looking upwards sees the hemisphere above the water as a solid angle of 97°. (After Lythgoe 1979.)

above. The dominance of downwelling light in aquatic environments profoundly affects the placement and structure of eyes and the shielding pigments in animals living there (Lythgoe 1979).

Animals on land and in the air have also developed a multitude of visual adaptations for hunting, predator avoidance and signalling in a variety of habitats where the directional distribution of light varies (e.g. open versus forested habitats). The major adaptations include modification of the direction and extent of the visual field, development of directionally and spatially sensitive areas of the retina, and binocularity. These mechanisms have been reviewed in detail by Walls (1942) and Hughes (1977).

3.3.6   *Characteristics of visual signals*

How do animals overcome the limitations just described, in order to communicate effectively by vision? Where light is scarce, bioluminescence is one solution. But many aquatic organisms appear to use bioluminescence for camouflage rather than signalling. Since the aquatic environment is dominated by downwelling light, prey organisms may have ventrally oriented photophores, the light output of which tends to match the background light from above and thus reduces the contrast that would otherwise make them visible to predators (Lythgoe 1979).

Fireflies, in which bioluminescence undoubtedly serves in courtship, communicate by flashes of light, or by glowing (Lloyd 1977; see section 2.3). Lall *et al.* (1980) recently showed that 21 of 23 species of firefly active at twilight produce yellow flashes with spectral peaks above about 560 nm. These signals almost certainly contrast effectively with the dim green light reflected from foliage; furthermore, the ambient light from the atmosphere at this time will be somewhat poorer in the yellow region of the spectrum because of the Chappuis effect (see Fig. 3.2a). Dark-active species of firefly tend to produce green flashes with spectral maxima below about 560 nm. Recall that moonlight is richer in longer wavelengths than daylight or twilight (Fig. 3.2a), so that the contrast of the flash against the background will be enhanced more than it would if the wavelength of the flash matched the dominant wavelengths of the ambient light. The spectral sensitivity of the visual receptors in these two classes of firefly tends to be 'tuned' to the spectral peaks of the flashes produced by their own species.

Visual communication in well-lit environments is efficient because the quantity and quality of light make possible the resolution of detail, movement and colour. Needless to say, visual displays are widespread and diverse, and the principal limitations are biological. Long-range visual signals are, almost by definition, conspicuous, but what is easy to detect and recognise by a conspecific is also obvious to a potential predator. Some visual signals can be turned on and off by movements that reveal or conceal a conspicuous colour or pattern. Indeed, sudden revelation of a striking colour or pattern may even startle a potential predator, and some permanently conspicuous patterns are common in distasteful or dangerous animals and their mimics.

Visual displays are conspicuous by virtue of colour and patterns of contrast such as boundaries, enhanced edges, discs, gratings and bars. These shapes presumably take advantage of visual mechanisms both at the receptor and higher levels. It is important to remember, however, that what may seem to be an obvious pattern in some environments may be highly inconspicuous in others.

3.4   Acoustic communication

Sound is a mechanical disturbance that propagates rapidly through air and water. Like pheromones, acoustic signals are effective by both day and night and in turbid or densely vegetated environments where vision is limited. Unlike chemical signals, sounds do not require a moving medium for effective transmission and, depending on their characteristics, sources of acoustic signals can also be localised easily. Sounds fade out quickly in most environments and thus can convey a great deal of temporal information.

Recent estimates of energetic costs indicate that sound production can be rather expensive. Of the singing insects which have been studied, the highest efficiency is estimated at about 10%, and a singing cicada may increase its metabolic rate as much as 20 times over that of a resting one (MacNally & Young 1981). In the frog *Physalaemus pustulosus*, males increase their metabolic rate by a factor of 4–6 when calling (Buchler *et al.* 1982). If sustained, energy expenditure on acoustic signalling can be staggering. MacNally (1981) estimates that only about 350 J of the 2600 J of energy assimilated by males of two small species of Australian frogs during a breeding season that lasts six months or more are converted to new growth; the rest is largely dissipated by calling.

3.4.1   *Environmental constraints on acoustic communication*

This section briefly reviews the technical literature on long-distance communication in various environments (see Piercy & Embelton 1977 for a technical review; Michelsen 1978 and Wiley & Richards 1978 also provide detailed surveys with biological examples). A definitive study does not exist because no researcher has measured simultaneously all of the relevant variables. Indeed, obtaining measurements of general applicability may be impossible because local microclimatic conditions and topographic features have important effects on the quality and quantity of sound propagation. Since sound travels as a pressure wave, albeit very different from an electromagnetic (light) wave, some of the same kinds of factors that affect visual communication are also relevant for acoustic communication, i.e. frequency- or wavelength-dependent absorption, scatter and boundary phenomena (reflection and refraction). This summary is restricted to sound communication in terrestrial environments.

Geometrical spreading

A point source of sound in a homogeneous medium radiates energy evenly in all directions; as the sound travels away from the source, the spherical volume of the sound field increases and thus the energy per unit area decreases. The relationship is simple: when the distance between the source and a receiver is doubled, the sound pressure level is halved. This inverse distance relationship is only approximately true very close to any real sound source. Furthermore, near-field (direct) displacement of air molecules (as contrasted with detection of pressure differences in the medium associated with a sound wave) may be detected by animals that communicate at very close range, such as some insects (Bennet-Clark 1971). Animals are not ideal point sources suspended in perfectly homogeneous environments, so real patterns of propagation may be cylindrical (giving less attenuation per doubling of distance) or loss-free (in rigid channels), or there may be a much greater loss (excess attenuation) than dictated by the inverse distance relationship (Michelsen 1978).

Atmospheric absorption

In still air, sound energy is lost as heat by various complex processes (see Piercy & Embelton 1977). High frequencies are more adversely affected than low ones, and this effect is increased as temperature rises but is moderated by high humidity. Other factors are usually more important than atmospheric attenuation.

Scattering and microclimatic effects

Scattering is a complex function of frequency (or, conversely, wavelength), the size, shape and rigidity of objects in the sound path, and heterogeneities in the air itself. If scattering objects are large and hard, total reflection may occur. Small objects, unless very much smaller than the wavelength of the sound, scatter or redirect the sound in complex ways that can augment or decrease sound levels, depending on the distance from the source and phase relationships of the direct and reflected or scattered sound waves. In practice, vegetation accounts for much of the scattering that occurs in forests. Temperature gradients are likely to form on a sunny day, especially in open habitats. One result is the formation of rising turbulent eddies of higher temperature than the surrounding air; these heterogeneities scatter and absorb sound in an irregular fashion.

Temperature gradients also give rise to shadow zones. The velocity of sound increases with temperature. Thus on clear days, when temperature decreases with height above the ground, sound refracts upward into zones of lower velocity. Shadow zones, which exist at some horizontal distance from the source (Fig. 3.4), are augmented by wind blowing in the same direction; the effect is moderated when sound propagation is upwind. At night, by contrast, cool air may be trapped near the ground and may provide a particularly good channel for sound propagation along the surface. In the early morning and at dawn, the dynamic properties of temperature-gradient formation or dissipation may even create channels for sound propagation that reduce losses to below the level expected by geometrical spreading (Wiley & Richards 1978).

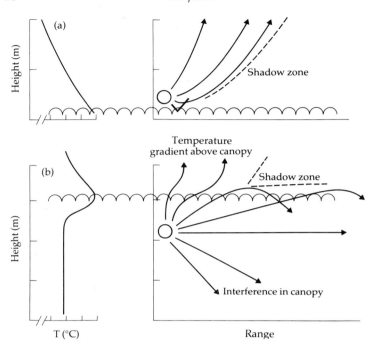

Fig. 3.4. (a) Diagram of shadow zone created by refraction of sound in the temperature gradient depicted at left. (b) Moderation of microclimatic effect (temperature gradient profile depicted at left) by the canopy of a forest. The diagram also depicts possible redirection of sound waves by reflection. (After Wiley & Richards 1978.)

Boundary phenomena

Animals usually produce signals near some boundary which reflects and absorbs sound. If both producer and receiver are near a reflecting surface such as water or asphalt, nearly all of the sound is reflected regardless of frequency. If, however, sound is propagated at a grazing angle to a porous substrate such as grass or loose soil, the reduction in high-frequency sound is much more pronounced than the reduction in low-frequency sound, even at a relatively short distance from the source. This effect is illustrated in Fig. 3.5b, which also shows the results of elevating the receiver to various heights above a grass substrate. The same partial recovery of high frequencies occurs if the receiver remains on the ground and the source is elevated. When both source and receiver are elevated the situation is much more complicated. Between the source and receiver there is now both a direct wave and one that is reflected from the ground. The direct and reflected waves interact constructively or destructively, depending on source–receiver geometry (the path-length difference between the direct and reflected waves), the wavelength of the sound, and changes in the phase of the reflected wave. If the frequency of the sound is held constant, sound pressure levels will rise and fall at various distances from the source. If the distance between the source and receiver is fixed, then sound levels will rise and fall as the frequency of the sound is varied (Fig. 3.5c).

3.4.2 Solutions to acoustic problems

The brief, simplified description above should give the reader an appreciation of the complexity of problems faced by animals attempting to communicate by sound. Although some general 'rules' emerge, many species apparently do not follow them for various reasons, some of them unknown.

In all habitats, low frequencies travel further than high frequencies, especially when both sender and receiver are very near a porous or vegetated substrate. Animals which, for anatomical reasons, cannot produce low-frequency sounds can improve broadcast efficiency by elevating themselves. This is well-illustrated by comparisons of estimated broadcast areas (area in which sound pressure level is above the behavioural threshold of the receiver) of ground-singing and arboreal-singing individuals of the cricket *Anurogryllus* (Paul & Walker 1979). Crickets that sang from trees or shrubs had broadcast areas that were at least 14 times greater than those on the ground, and males singing at heights of 0.5 to 2.0 m attracted more females than males singing from higher elevations or at ground level (except at the burrow).

Other solutions are illustrated by mole crickets and some Australian species of frog. These animals construct burrows with particular dimensions that enhance the efficiency of their sound production (Bennet-Clark 1970; Bailey & Roberts 1981). A tree cricket actually uses leaves to build a sound baffle for the same purpose (Prozesky-Schulze *et al.* 1975).

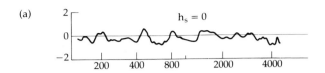

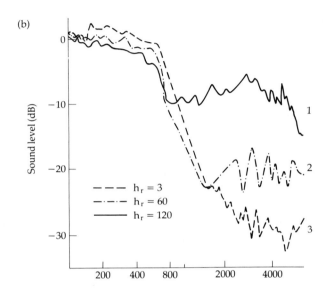

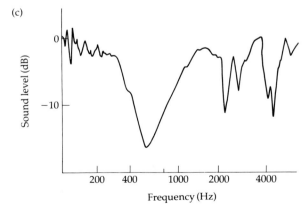

There is a definite tendency for tropical forest birds to sing at lower frequencies than do birds in adjacent open habitats. Morton (1975) proposed that this is a consequence of the fact that low frequencies propagate more effectively in forests than do high frequencies. However, the same correlation between habitat and pitch of song is not seen among temperate birds in North Carolina (Wiley & Richards 1983). Wiley and Richards argue that one reason for this discrepancy is the fact that most birds in temperate forests sing from elevated perches, whereas those in tropical forests mostly call from close to the ground. Communication near the ground clearly limits high-frequency signals. Some confusion still exists, however, because ornithologists and entomologists understandably have different ideas about what 'close to the ground' means. Elevation of a source and receiver by just 0.5–2 m can cause significant destructive interference of frequencies around 500–1000Hz (relatively low as far as birds are concerned) over the distances (about 100 m) by which singing birds are often separated (Marten & Marler 1977). Recall, however, that such boundary effects are complex functions of frequency and exact source–receiver geometry (Fig. 3.5): limitations that may apply to a bird singing at 1000 Hz to a rival at 100 m may not apply to a cricket singing at 4000 Hz to another individual 10 m away.

Another, more interesting, difference between the songs of forest birds and those of open fields is their time pattern or temporal structure. Morton (1975) found that birds living in forests tended to produce monotonous tonal calls that changed rather

---

Fig. 3.5. Propagation of sound as a function of frequency over a horizontal distance of 15.3 m. The zero reference point in each diagram is attenuation expected by geometric spreading (see the text). (a) The source is near a hard, reflecting substrate (asphalt) and the receiver is elevated 1.2 m above the same substrate. Notice that frequencies up to at least 4 kHz propagate equally well. Similar results are obtained when both source and receiver are near such a hard, reflecting surface. (b) The source is near a grassy substrate. If the receiver is also near the substrate, there is severe excess attenuation of the high frequencies (curves 2 and 3). Elevation of the receiver (h_r = height of receiver in centimetres) moderates this high-frequency attenuation (curve 1). The relationship is reciprocal. Elevating the source as the receiver remains at ground level results in a similar recovery of high-frequency energy at the receiver's position. (c) Both the source and receiver are elevated over a grassy substrate. Notice that destructive interference reduces sound level at about 500 Hz, 2000 Hz and 3800 Hz. This pattern is specific for the particular source–receiver geometry (heights and horizontal distance). (After Embleton *et al.* 1976.)

slowly in frequency, whereas birds in open habitats more often produced rapid trills, or bursts of sound that change rapidly in amplitude and frequency. Wiley and Richards (1978) found a similar, though less pronounced, tendency in the temperate zone and proposed that a primary factor in this is the different kinds of scatter that predominate in the two habitats. In forests, scatter mainly takes the form of reverberations caused by the interaction of the sound with vegetation and other objects. Reverberations effectively mask or blur the discontinuities in a rapidly modulated song, thereby distorting its temporal pattern. By contrast, the dominant form of scatter in open areas is caused by air turbulence arising from temperature gradients or wind. The resulting interference takes the form of random fluctuations at low rates (less than 50 Hz), uncorrelated with the occurrence of the song. Thus, rapid modulations of amplitude and frequency in a song may be desirable since they will only intermittently be degraded. Enough of the temporal pattern should be transmitted clearly for recognition to occur.

Still another strategy is to produce broad-band sounds, with many widely separated frequency components. In situations where frequency-dependent losses are severe and unpredictable, at least some of the sound pattern should get through. The disadvantage is that such a signal probably will not excite the auditory apparatus as effectively as a pure, tonal signal, where all of the energy is concentrated in a narrow band. Thus the maximum range of transmission could be limited.

Since microclimatic factors are so important, there are optimum times to signal. At dusk and dawn, propagation losses may be dramatically reduced, especially in forests where the canopy forms a physical barrier and microclimatic buffer (see Fig. 3.4; Wiley & Richards 1978). Conditions at night are also favourable, and there is the additional advantage that many predators are diurnal. The prevalence of dusk/dawn choruses and nocturnal sound producers supports these ideas (Henwood & Fabrick 1979; Young 1981). Midday is probably the least favourable time to signal, especially on sunny days in open habitats. Recall that temperature gradients cause shadow zones and rising pockets of turbulence even on calm days.

Although most authors have been preoccupied with estimating maximum transmission distance through various habitats, this may not be the major concern of many animals. A great deal of signalling occurs at relatively close range where, even if visual contact is restricted, an acoustic signal is clearly audible. Wiley and Richards (1978, 1982) suggest that animals could use signals that have some degradable properties for estimating the distance to the source. In the view of the author, acoustic environments are too complicated and variable for accurate range assessment, but properties that degrade differentially with distance probably provide the receiver with a spatial sequence of cues, rather like a pheromone plume consisting of a mixture of different compounds. The green treefrog (*Hyla cinerea*), for example, produces a complex call with two major frequency bands and a distinctive temporal structure. Females are attracted at very low sound levels (corresponding to a long distance) by just the low-frequency peak or even by a continuous low-frequency noise band which lacks a temporal structure (Gerhardt 1976; Ehret & Gerhardt 1980). As a female gets closer to a male and the sound level increases, she selectively responds to calls with both low- and high-frequency bands, and rather subtle temporal differences also come into play (Gerhardt 1978b).

3.4.3   *Sound localisation*

In general, sounds are intermediate in locatability between chemical and visual signals. Marler (1959) proposed a set of variable signal properties which affect the ease with which an animal can find the source of the signal. He reasoned that sounds with sharp onsets or of very low frequencies provide accurate time cues for comparisons of external time of arrival or phase between the two ears of a predator or conspecific. When an animal orients so that a sound arrives at both ears simultaneously, for example, it faces directly towards or away from the source. Broad-band or high-frequency sounds facilitate localisation by comparison of interaural intensity and spectral (complex frequency pattern) cues. Sound scattering (shadowing) by the head and external ears (in mammals) accentuates external intensity differences; complex frequency spectra presumably also allow the animal to tell whether the sound is coming from directly ahead or behind. Signals used for intraspecific communication or for some kinds of interspecific communication (e.g. mobbing in birds), where accurate localisation is important, incorporate many of these properties. Warning or alarm signals, by contrast, are usually tonal, have gradual onsets, and are too high in pitch for phase comparisons but too low in pitch for effective sound shadowing. These features would presumably make them difficult for predators to locate (see section 2.9.1).

Marler's arguments presumed that the ears of animals function as pressure receivers, and there is a great deal of evidence that higher mammals use this mechanism. In such a system, sound pressure changes stimulate the ears only on their external surfaces (tympanic membranes). Recent studies indicate, however, that many, if not most, other kinds of vertebrates may use a very different mechanism for sound localisation, a mechanism that is well-known in insects (Lewis & Coles 1980). In these animals the ears seem to function as pressure-gradient receivers. Binaural effects are still important, but they are generated by an interaction between two sound waves, one of which travels through internal pathways within the animal and activates the ear (tympanic membrane) from the inside. Many birds, reptiles and amphibians have ears that are close together and communicate by sounds that have long wavelengths relative to the dimensions of their heads. This means that head shadowing is inadequate for creation of differences in intensity at the two ears and that interaural time-of-arrival differences are minute. Shalter (1978) found that small predatory birds appear to localise the alarm calls of the small birds upon which they feed as accurately as they could localise other sounds which should be much more easily pinpointed, but Brown's (1982) study of large hawks and owls confirmed Marler's predictions. Brown speculates that alarm call frequencies are also difficult to localise by a pressure-gradient receiver with dimensions dictated by the head size of these larger predators.

A pressure-gradient system has an advantage over a pressure-receiver system in that low frequencies as well as high frequencies can be localised. Studies of the green treefrog (*Hyla cinerea*), for example, indicate that females can accurately localise synthetic male calls consisting of just the low-frequency components (Rheinlaender *et al.* 1979). Since low frequencies propagate better than high frequencies, the distance over which the animals can localise one another is greater than it would be if the high-frequency components were necessary.

3.5 Noise in animal communication

An organism is constantly bombarded by sensory stimuli, and unless its receptors are very narrowly tuned, irrelevant stimuli as well as important signals will activate its receptors. Narrow tuning greatly reduces the flexibility of communication because the sounds produced by mates, rivals, predators and prey are unlikely all to have the same frequency. Thus many animals must cope in some way with noise. The most general effect of high background noise is the elevation of the threshold for signal detection and recognition. One solution, adopted by some animals, is to increase the repetition of signalling (see Chapter 5). The receiver can thus average over time to increase the chances of detecting and recognising the signal.

In chemical communication systems, noise appears to be primarily biological, and the problem of species-specificity in pheromone recognition has already been discussed (section 3.2.3). In visual systems, background 'noise' consists mainly of reflections and scatter from vegetation, rocks and the substrate, and of numerous factors that affect light quantity and quality in water. Most of these factors reduce the contrast which is necessary to detect and recognise a potential signal. Species-specific colour and contrast patterns and display movements generally become less effective at a distance or where ambient light is limited.

We more commonly associate the term 'noise' with sound because of its everyday meaning. The principal natural, non-biological sources of noise are running or falling water, wind, and sounds made by wind-blown vegetation. These sounds contain a broad band of frequencies, and there is no easy solution for an animal other than to avoid communication at unfavourable times and places, or to combine acoustic signals with signals of other modalities.

Sources of biological noise are often very impressive, and mixed choruses of birds, insects and frogs can include dozens of species. The potential interference caused by so many animals broadcasting at the same time is one of the drawbacks of restricting sound production to favourable times (see above). Although there may be effective separation in terms of broadcast frequency, such as in some Puerto Rican frog choruses (Drewry & Rand, in preparation), frequency differentiation in most regions is not so obvious. Dominant frequencies of frog calls tend to cluster between about 1500 and 3000 Hz, and choruses often include two or more species calling at nearly the same frequency (Loftus-Hills & Johnstone 1970). The major limitations are probably anatomical constraints on the frequencies that can be produced.

One solution is to partition broadcast time, and this device is used by many animals which communicate by sound. Indeed, the calls of some frog species effectively inhibit other species from calling; the inhibited species calls during pauses in the vocal activity of the dominant species (Littlejohn & Martin 1969; Schwartz & Wells 1983). Another solution is to decrease auditory sensitivity and to tune the receptor to the main communication frequency, as in the North American cricket frog (*Acris crepitans*) (Capranica *et al.* 1973). Although this procedure limits the range of communication and, as already mentioned, flexibility, it effectively reduces interference caused by biological noise. Finally, it is becoming clear that some patterns of signalling may be designed to interfere with signalling by conspecific competitors. In some Panamanian treefrogs, males add secondary notes to their calls and these tend to disrupt the calling patterns of their neighbours (Wells 1980). This phenomenon also has parallels in the flash communication systems of some species of firefly (Lloyd 1981).

3.6   Special problems of communication and the environment

3.6.1   *Temperature effects on communication in cold-blooded animals*

Insects and lower vertebrates may signal over a fairly wide range of temperatures. Some properties of their signals, especially temporal ones, may thus change in significant ways because of the effects of temperature on metabolic rates and neuromuscular mechanisms. Folklore correctly asserts that temperature can be estimated by listening to crickets. These changes in signal properties raise an interesting question. Does a receiver that identifies signals by their temperature-dependent characteristics take into account its own temperature or that of the environment?

Two species of gray treefrog, *Hyla versicolor* and *H. chrysoscelis,* are widespread in the eastern half of North America, often

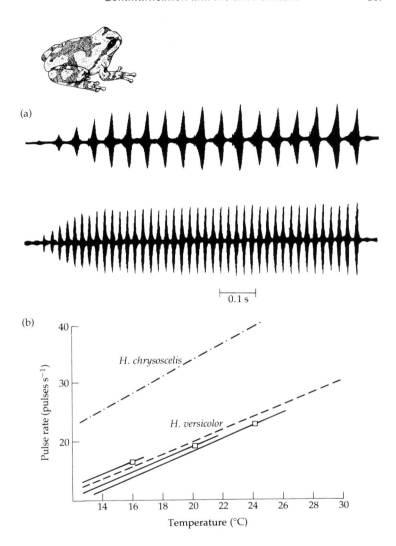

Fig. 3.6. (a) Oscillograms showing the trills of the gray treefrogs *Hyla versicolor* and *H. chrysoscelis* recorded at about 18°C. The pulse-repetition or trill rate is the reciprocal of the time from the beginning of one pulse to the beginning of the next pulse. (b) Pulse-repetition rate of the calls of the gray treefrogs as a function of temperature. Notice that a 'warm' *H. versicolor* produces calls with a trill rate which is similar to that of a 'cool' *H. chrysoscelis*.

breed at the same time and place, and produce calls of about the same frequency. At any one temperature, the pulse-repetition rates (trill rate) of the calls of the two species differ significantly, but the trill rate in *H. versicolor* at high temperatures (> 25°C), is similar to that of *H. chrysoscelis* at low temperatures (< 16°C) (Fig. 3.6). Thus at low temperatures if a female of *H. versicolor,* which recognises the male solely by his species-specific call, did not take her own temperature into account but merely responded to any call with a trill rate within her species' range of variation, she would risk mating with a male of *H. chrysoscelis* calling at low temperatures. Similarly, a female of *H. chrysoscelis* would risk mismating at high temperatures. Mismating is one of the most serious blunders that a female can make; in this particular example, the results are sterile hybrid offspring.

Playback experiments with both species indicate that female preferences for calls with different pulse-repetition rates do change with temperature. Furthermore, these changes in preference are just those the species identification problem dictates. Females of *H. versicolor* discriminate very strongly against calls with higher trill rates (in the direction of the other species), especially when tested at low temperatures. Females of *H. chrysoscelis* discriminate strongly against calls with lower trill rates, especially at high temperatures. Increasing the temperature of a female of *H. versicolor* makes calls with higher trill rates more attractive; decreasing the temperature of a female of *H. chrysoscelis* makes calls with lower pulse-repetition rates more acceptable (Gerhardt 1978a, 1982). Temperature-dependent changes also occur in the temporal properties of insect songs, flash patterns of fireflies and the electric discharges of fish, and these changes appear to be paralleled by temperature-dependent changes in the preferences of receivers, a phenomenon called 'temperature coupling' (Gerhardt 1978a).

Temperature coupling is not always the rule. Unlike trill rates, the frequencies of frog calls change relatively little with temperature. In at least one species, the green treefrog (*Hyla cinerea*), a decrease in temperature lowers the preferred frequency of the female (Gerhardt & Mudry 1980) but has little effect on the frequency of the male's call. The change appears to depend on temperature effects on the auditory system, and other frogs have tuning mechanisms similarly affected by temperature (Walkowiak

1980). This potential problem in communication may be avoided by restricting communication to temperatures at which receptor sensitivity matches the broadcast frequency of the calls.

3.6.2 *Exploitation of communication systems by predators*

Signals must often be conspicuous for effective transmission in natural environments, especially noisy ones. The most obvious liability is that predators may use the signals to localise the signaller. This almost certainly occurs in all communication modalities. Particularly well-documented examples involve singing crickets and calling frogs. Cade (1979) found that singing males of *Gryllus integer* were more often victimised by a parasitoid fly, *Ephasiopteryx ochracea,* than were silent ones. The tropical frog *Physalaemus pustulosus* has a wide variety of predators that appear to locate calling males at least in part by their voice: a large toad, *Bufo marinus;* the four-eyed opossum, *Philander opposum;* and the fringe-lipped bat, *Trachops cirrhosus* (Ryan *et al.* 1981). Predation by bats is particularly impressive: 95 frogs were eaten in about 14 hours of observation. Males of *Physalaemus* face another dilemma: the call type (a more complex one) that best attracts females is also preferentially approached by the bat. Ryan *et al.* argue that individual risk is reduced by calling in large aggregations and that more females are also attracted to these compared with small choruses. Except on very dark nights, calling males detect bats effectively by vision, and stop calling. Cessation of calling by one or two males is apparently perceived by the others, and the chorus shuts down immediately. Even more impressive is the fact that males differentiate between the predatory bats and smaller, insectivorous species; the frogs usually ignore the small bats and continue to call (Tuttle *et al.* 1982).

In the crickets studied by Cade, and in the green treefrog and other frogs, silent males position themselves near calling ones and often intercept females that are attracted by the caller (Cade 1979; Perrill *et al.* 1978). These so-called satellite males not only reduce their risk of predation, but also avoid the considerable energetic costs of calling (see section 3.4). It is surprising that satellite males are only rarely observed in *Physalaemus,* where the costs of calling are so great.

A more interesting form of exploitation of communication

systems by predators involves mimicking the signals of another species. Rather than finding a mate, responsive individuals become a meal for the predator. One species of spider in New Guinea actually produces a pheromone that attracts male moths into its web (Eberhard 1977).

The most fascinating and detailed examples of this form of exploitation come from the predatory fireflies of the genus *Photuris* studied by Lloyd (1981). Male fireflies of many different species patrol for mates while emitting flashes of light. Species-specificity resides in the number and timing of flashes as well as their wavelength, duration and the temporal pattern of intensity within each flash. When a responsive female recognises the flash pattern of a conspecific male, she answers with a flash pattern of her own. This pattern not only has a species-specific form, but its timing relative to the flashes of the male is highly specific. Both the timing of the male's flashes and the delay of the female's answer are affected by temperature. On any given evening there are many more patrolling males than responsive females, and so sexual competition is intense. A female's answer elicits a rapid response. Lloyd (1981) found that a female of the firefly *Photinus collustrans* can emerge from her burrow, answer a male, attract him to her, copulate, and return to her burrow in as short a time as six minutes.

The strong competition for mates makes the males of some species fairly reliable prey for predatory females of *Photuris*. In *Photuris versicolor*, for example, a female finds an area where prey species are active and settles near the ground. She recognises the prey's flash pattern and answers with flashes that resemble those of the prey species female in both form and timing (Fig. 3.7). *Photuris versicolor* can attract males of at least five other species of firefly, and Lloyd's studies of four prey species indicate that about 16% of the attracted males are trapped and eaten.

Naturally this form of predation has affected the behaviour of the prey. The patrolling male may not fly directly to the answering flash, but will approach and withdraw repeatedly, or even fly away. Of course, if he hesitates he may lose a rare opportunity to mate. Males of some prey species, such as *Photinus collustrans*, appear to be particularly adept at avoiding capture. Although they usually reached conspecific females in less than a minute, none of the 11 males that Lloyd observed to be signalled to by *Photuris*

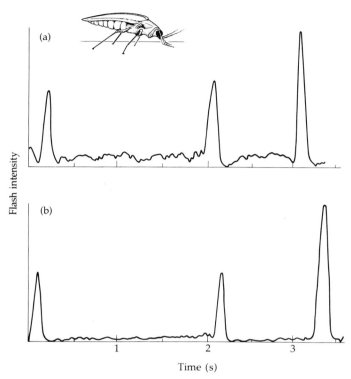

Fig. 3.7. (a) Flash pattern of a male of *Photinus macdermotti* (two flashes of the same intensity) and of an answering female (the single more intense flash). (b) Imitation of the twin-flash pattern of the male by Lloyd (1981), and answering flash of a mimicking *Photuris versicolor*. Based on photomultiplier field recordings. (Adapted from Lloyd 1981.)

versicolor approached close enough to be seized. Lloyd (1981) suggests that the males may be able to perceive subtle discrepancies in the mimic's flashes or may actually identify her in the twilight by her large size. Alternatively, males may alter their signals in order to lead the predator to give a response with faulty timing, different from that of a conspecific female.

A remarkable consequence of mimicry by these females is the evolution of males that mimic the flash patterns of one or two of the prey species (Lloyd 1980, 1981). Males of *Photuris* usually adopt the same habitat, altitude, and seasonal and daily activity periods as the mimicked species; unlike *Photuris* females, however, they probably do not use deceit to obtain their food. Rather, it appears that males are attempting to locate and mate with hunting (mimicking) females of their own species. Lloyd (1981) observed that mimicking males switched to the conspecific (non-mimetic) flash pattern after landing near answering females.

3.7   Conclusions

Chemical communication can be used by both day and night and in a wide variety of habitats. The signals may be brief in duration or long-lasting. Effective distances of communication depend on wind or water current, the flow rate of which must be neither too slow nor too fast. Localisation of pheromones is often uncertain; again, a requirement for long-distance orientation is an appropriate wind or current. Chemical communication systems may be so specific that identification of signals is assured even when other species are also using similar pheromones, but examples of cross-species attraction are not rare.

The quantity and quality of ambient light constrain visual communication, except in bioluminescent species in which the spectrum of the light emission may be selected for maximum contrast with the photic background. Aquatic organisms must deal with the strong filtering of long-wavelength light through water, and terrestrial creatures with visual veils created by scatter. Shifts in absorption maxima of visual photopigments and special filters are solutions used by some animals. Conspicuous patterns or movements that contrast with the background are necessary for long-distance communication, but have the disadvantage of attracting predators. Patterns that appear conspicuous to humans against some backgrounds may be very cryptic in natural habitats. Localisation is seldom a problem because of the directional properties of light and of most visual receptors.

Acoustic signals can travel rapidly over fairly long distances, and can be adapted for ease or difficulty of localisation. Like pheromones, sounds are effective in a variety of habitats during most of the daily cycle, but are especially effective at night and during day–night transitions. The frequency and temporal properties of signals can be related only crudely to habitat differences because of the enormous variability of relevant factors. These include atmospheric absorption, scattering and random amplitude fluctuations caused by vegetation and atmospheric heterogeneities, boundary effects, and turbulence and shadowing caused by microclimatic phenomena. Some general 'rules' emerge, but limitations of anatomy and other factors constrain the design of animal signals. These problems are compounded by biological sources of noise; temporal partitioning of broadcasting and the use of different frequency bands are two common devices to combat such noise.

Poikilothermic animals often signal over a wide range of temperatures, and some signal properties consequently change dramatically. In some species of crickets, fireflies, frogs and electric fishes changes in temperature also appear to shift the preference or sensitivity of receivers in a roughly parallel fashion.

Predators exploit communication systems of other species by using the prey's signals for identification and localisation. Prey may detect the predator and stop calling. Conspecifics also cheat signallers, by remaining silent and intercepting mates. Another predator strategy is to mimic sexual signals of prey species so that responding individuals become victims instead of mates.

3.8   Selected reading

An excellent introduction to olfactory communication is *Animal Communication by Pheromones* by Shorey (1976). This deals especially with the behavioural effects of pheromones and has an exhaustive bibliography. Detailed accounts of animal visual systems and of the properties of photic environments can be found in *The Ecology of Vision* by Lythgoe (1979), a fully illustrated book with an extensive bibliography. An excellent introduction to acoustic communication is provided by Wiley and Richards (1978). This paper gives a detailed, though rather technical, account of auditory signals in natural environments with a wealth of biological examples and a review of the recent literature. The complexities of interspecific exploitation in a communication system are explored in a highly recommended paper by Lloyd (1981) on fireflies.

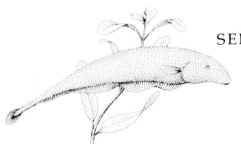

CHAPTER 4
SENSORY MECHANISMS
IN ANIMAL
COMMUNICATION

CARL D. HOPKINS

4.1   Introduction

The description and analysis of animal communication has always presented a challenge to ethologists because there have to be at least two participants involved, a signal sender and a signal receiver, whose actions must be followed simultaneously. The actions of the sender are often easier to describe than those of the receiver because display actions tend to be stereotyped or fixed in their form, conspicuous and easily recognisable, and sometimes repetitive. Responses to signals, by contrast, tend to be variable in form, probabilistic in occurrence, strongly modified by the context in which the signal occurs, and delayed by varying amounts of time after the occurrence of the signal. It is not surprising, therefore, that the literature on animal communication has tended to emphasise the description and analysis of signals rather than of the responses to them.

Clearly, the mechanisms of signal reception and the ensuing responses are under the influence of natural selection, as are the forms of the signals themselves. The receiver attends to and is aroused by the signal, it detects and filters the signal in the presence of noise, it discriminates among different types of signals, and it locates the signal in space. The purpose of this chapter is to direct attention to the receiver's viewpoint in the communication process, and to examine the mechanisms of signal reception and their adaptive significance. Concern with the nature and evolution of signal reception leads inevitably to the large and complex discipline of sensory physiology, but this is beyond the scope of the present chapter. It is possible, however, to look in some detail at the interface between ethology and sensory physiology, as there has been considerable work on the sensory processing of biologically significant signals (see Ewert *et al.*, in press; Guthrie 1980).

To understand how a sensory system used in animal communication is adaptive, we must seek a theoretical framework for predicting optimal sensory design. Communication engineers might be able to provide such a framework, if we could give them sufficient information about the nature of signals and the properties of the communication channel. One goal of this chapter, then, will be to compare the designs of different communication sensors using different sensory modalities, with those expected from basic communication engineering principles.

The engineer who is faced with the task of designing a communication receiver, after first selecting a receiving transducer to match the physical nature of the stimulus (the sensory modality), will have to decide upon the appropriate sensitivity and selectivity of the transducer or receiver. By specifying the sensitivity of the receiver, the engineer determines the minimum amplitude of the signal which can still be received, and hence the range of signal reception, given a constant signal amplitude. The selectivity of the receiver will affect signal detection in the presence of jamming or interference in the communication channel. Filters must be built to eliminate interference and improve signal-to-noise ratio. The most difficult task will be the design of mechanisms to discriminate amongst desired signals and to respond differentially to them. One receiver may be highly selective, and reject all but a few types of signals, and another may accept a wide variety of signal types. The design may incorporate mechanisms to insure that the signal can be localised in space. At every stage of the design, of course, the engineer will be attempting to reduce the overall energy consumption of the receiver, as well as its cost, weight and size. The parallels with the evolution of a communication sensor are obvious.

Four related design considerations will be discussed in this chapter: first, the means by which the recipient improves its chances of receiving the signal when it is present (and conversely, not detecting it when it is absent), even when the signal is buried under other unwanted stimuli (*filtering*); secondly, how the signal gets translated into an internal message by the recipient (*coding*); thirdly, how the recipient recognises the signal of interest, and takes appropriate action (*recognition*); and fourthly, how the origin of the signal in space is detected (*localisation*).

4.2   Receptors for communication—four examples

We may reasonably ask to what extent sensory structures have evolved to receive signals. Are there specialisations, for example, that permit the receiver to filter specifically for one class of signal, to discriminate amongst some signals but not others, or to localise certain types of signals in space? Such questions have received increasing attention since sensory physiologists began to examine the way certain sensory neurons respond to certain key features of stimuli (see Lettvin *et al.* 1959). In fact, some animals have sensory systems the sole function of which appears to be the sensing of communication signals; the sensors are not used for other, more general sensory tasks. Four examples of such 'intraspecific communication sensors' will be discussed in this introductory section. These examples may be exceptions, rather than following the general rule, but they stress the point that receiving a communication signal is a highly evolved and specialised sensory function. The four examples are drawn respectively from olfaction, audition, electroreception and vision. In each case, the description of the behaviour precedes that of the relevant anatomy and physiology.

4.2.1   *Sex attraction and olfaction in insects*

The first example comes from the often-cited olfactory communication sensors of nocturnal moths like *Telea polyphemus,* or *Bombyx mori,* the silk-worm moth. The olfactory receptors appear to be intraspecific communication sensors because they are sexually dimorphic, matched to the sex-related difference in attraction behaviour, because the receptors are such specialised filters for incoming olfactory stimuli, and because they have a low threshold for species-typical odours. In these and other moths, females release an attractant odour (pheromone) from their abdomens and males respond to the odour by flying upwind towards its source (reviews by Schneider (1965, 1969) and Kaissling (1971)). Communication is sex-specific and is unidirectional, in that females act only as signallers and males as receivers. The sensor is correspondingly sexually dimorphic, especially in the saturnid moths. In one saturnid, *Antheraea pernyi*, for example, the outline area of the male antenna is 55 mm^2 compared to 10 mm^2 for the female; in *Bombyx* the female's antenna is 5.5 mm^2 compared to the male's 6.0 mm^2.

The olfactory sensors on the moth antenna are the long, hair-like *sensilla trichodea*, and the shorter *sensilla basiconica*, which are fluid-filled extensions of the cuticle, into which one or more neurons send distal dendrites. The cuticle is perforated with 100–150 Å diameter pores which serve as channels for odour molecules. In *Bombyx* males, there are over 16 000 of these sensory hairs on the antenna surface, each aligned to maximise contact with airflow over the antenna. Antennae of some moths like *Bombyx* are highly efficient at capturing pheromone molecules. Part of the reason stems from the large cross-sectional area of the antenna already mentioned. To get quantitative measures of the affinity of the antenna for pheromone, Kaissling and Priesner (1970) passed radioactively labelled synthetic pheromone bombykol (hexadecadiene-10-*trans-cis*-12-1-ol) over the antenna and found that 27% of the molecules in the airstream were adsorbed. Of these, 75% reached active sensory hairs. Kaissling and Priesner calculated that 650 molecules per cubic centimetre are needed to trigger a behavioural response. Some males respond to a female's odour at distances as great as 1 km.

Insect antennae are designed not only for sensitivity but also for selectivity. To describe selectivity in terms of odours involves the description of a reaction spectrum, i.e. the measurement of thresholds to a broad spectrum of molecules. This has been done for some insect species, and the work on *Bombyx* by Schneider and his colleagues is among the most complete. Schneider *et al.* (1967), recording from single units in the antenna of *Bombyx* stimulated with stereoisomers of the synthetic pheromone (*cis-cis-, trans-trans-, cis-trans-*, etc.), found that these slight modifications made the molecules 100 to 1000 times less effective. They concluded that the *sensilla trichodea* are specialist sensors, responding to only a few compounds at low concentration.

In these moths selection obviously favours the most sensitive and selective male recipients. While the females broadcast their odours widely, the males that receive the signals at the greatest distance, and that make fewest mistakes in recognition, will be most successful at finding mates. Mate detectors using chemical systems are selected for efficient filtering, high sensitivity and accurate recognition.

4.2.2   Sex differences in frog audition

A second example of a communication sensor comes from a treefrog in Puerto Rico, *Eleutherodactylus coqui*. Male and female frogs differ in their auditory systems in ways that appear to be adapted to whether they need to sense sounds used in male–male communication or those used in communication between the

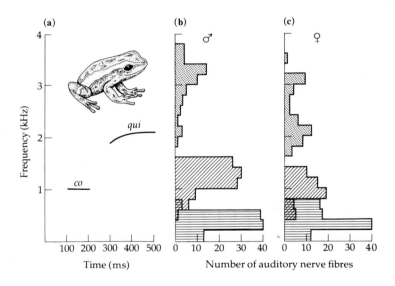

Fig. 4.1. Male frogs from Puerto Rico (*Eleutherodactylus coqui*) produce a call with two parts: the 'co' note which is used in intra-male aggression, and the 'qui' note, used for calling females. The two are given in the pattern shown in the sound spectrogram (**a**). Males echo synthetic 'co' notes with 'co' notes of their own. Females approach synthetic 'co-qui' calls, but not the 'co' call alone. Male and female frog auditory nerve fibres are studied electrophysiologically (**b** and **c**), and optimal frequencies of single auditory units fall into three classes dependent upon frequency. Male and female mid-frequency units are both matched to the frequencies of the 'co' call. The female's high-frequency receptors are matched to the 'qui' note, but the male's high-frequency receptors are tuned too high to match the 'qui' note. (Adapted from Narins & Capranica 1976.)

sexes. During the breeding season, males produce a mating call consisting of two notes: an unmodulated pure tone 'co' note at about 1000 Hz, followed by an upward-modulated 'qui' note spanning the range of frequencies from 1800 to 2200 Hz (Fig. 4.1a). Males and females react differently upon hearing the call. If a resident calling male hears a nearby frog give a 'co-qui', he will usually respond quickly with the single 'co' note; he may later attack the intruding male. Narins and Capranica (1976) played synthetic calls to resident male frogs and elicited single 'co' notes in nearly half of their trials. When the 'co' note alone was played to males, they echoed with their own 'co' responses; the 'qui' note alone provoked the males to respond with the full 'co-qui'. Females, on the other hand, approached loudspeakers producing either 'co-qui' calls or just 'qui' notes, but failed to react to the 'co' note by itself. One can conclude that the two notes serve distinct functions: the 'co' is used in inter-male communication, primarily in territorial defence, and the 'qui' note is used in mate attraction.

By investigating the auditory system of *Eleutherodactylus coqui,* Narins and Capranica were able to demonstrate sex differences in the distribution of tuned auditory nerve fibres (Fig. 4.1b, c). Like many of the frog species that have been studied, *E. coqui* has three types of auditory units, sensitive to low, mid and high frequencies (100–600 Hz; 1400 Hz; 1800–3700 Hz). Both males and females have an abundance of mid-frequency units probably originating in the amphibian papilla, one of two auditory organs found in frogs (Feng *et al.* 1975). These units are highly responsive to the frequencies in a typical 'co' note. Females also have high-frequency units tuned to the 'qui' note, but high-frequency units in males are tuned too high to be optimal receivers of the 'qui' note (see Fig. 4.1b). The high-frequency units probably originate in hair cells on the second auditory organ, the basilar papilla, which is believed to be homologous to the mammalian cochlea. Females are thus specialised for receiving the male's mating call, specifically the 'qui' note, and males appear specialised for hearing intra-male territorial calls, the 'co' notes. Since it appears unlikely that differences in ecological conditions between males and females can account for the sex differences in the auditory system, it is likely that these differences arise to meet communicatory functions.

(a)

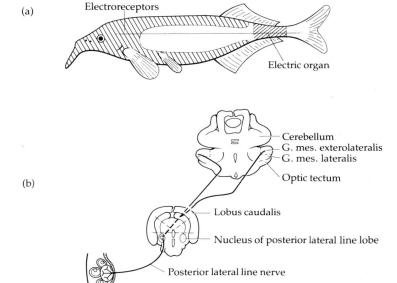

Electroreceptors

Electric organ

(b)

Cerebellum
G. mes. exterolateralis
G. mes. lateralis
Optic tectum

Lobus caudalis

Nucleus of posterior lateral line lobe

Posterior lateral line nerve

Knollenorgan

(c)

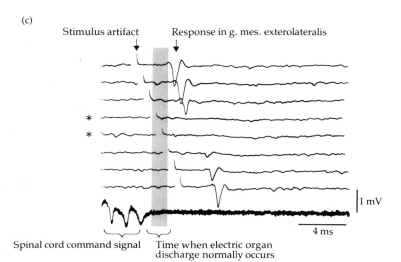

Stimulus artifact   Response in g. mes. exterolateralis

*

*

1 mV

4 ms

Spinal cord command signal   Time when electric organ
discharge normally occurs

4.2.3   Electroreceptors for communication

The third example of a communication sensor comes from the African electric fishes of the family Mormyridae, which possess electroreceptors known as Knollenorgans. Mormyrids have a well-developed system of social communication using electric signals; they produce electric displays in the context of aggression, during courtship, and for promoting aggregation and dispersal. Electric communication is important in sex and species recognition (reviews in Hopkins 1974, 1977). In all, mormyrids have three types of electroreceptors distributed over the skin surface: ampullary organs, mormyromasts and Knollenorgans (Szabo 1974; Bennett 1971). There are three corresponding functions for electroreceptors. The first is passive electrolocation, in which the fish senses a steady D.C. field given off by a predator or prey organism. The second is active electrolocation, which is analogous to echolocation in bats, where the fish produces an electric field in the electric organ in its tail, senses the distortions in the field surrounding its body, and thereby 'feels' the presence of distant objects with conductivities different from that of the water. The third function is communication, where one fish senses the electric signals of a second, using its electroreceptors.

There are three reasons to believe that Knollenorgans have evolved to receive communication signals. First, they are quite sensitive as receptors, responding to electric fields of as little as

Fig. 4.2. Mormyrid electric fish *Gnathonemus petersii* (a) possess Knollenorgan electroreceptors whose putative function is social communication. Electroreceptors lie in a specialised epidermis (front shaded zone). The electric organ is found in the caudal peduncle. (b) The Knollenorgan receptor projects to the ganglion exterolateralis of the midbrain. Electrophysiological recordings there reveal the effects of the correlary discharge to fire the electric organ on the input from Knollenorgans. G. mes. exterolateralis = ganglion mesencephali exterolateralis, G. mes. lateralis = geniculus mesencephali lateralis. (c) A brief electrical stimulus is applied across the skin, just sufficient to excite Knollenorgan receptors. The response recorded extracellularly in the midbrain arrives after a 3–4 ms delay. Responses to stimuli which are delivered 2.5 to 6 ms after the command signal to fire the animal's own electric organ arrive in the tail of a curarised (electrically silent) fish (bottom trace). The sensory input to the midbrain is blanked (traces marked *) at the precise moment when the Knollenorgans expect to relay input from the fish's own discharge. Hence the fish never hears its own discharge with its Knollenorgans! (b adapted from Szabo *et al.* 1979; c adapted from Russell & Bell 1978.)

0.1 mV cm⁻¹ (Bennett 1971; Hopkins 1981), a field strength equivalent to that set up by a dry-cell battery connected to electrodes at
the ends of an Olympic swimming pool! Because electric fields
attenuate rapidly in water (Knudsen 1975; Hopkins 1974), this
acute sensitivity will be crucial to sensing of communication
signals. Secondly, the frequency sensitivity curve of the Knollenorgan receptor fits the spectrum of energy in the electric discharge,
so that there is a type of spectrum matching of species-specific
signal output with receptor input (Hopkins 1981). Thirdly, and
most importantly, mormyrids appear to be unable to hear their own
output through the central pathway connected to the Knollenorgan receptors because of an inhibitory interaction between the
command that fires the electric organ and the sensory input from
Knollenorgans (see Fig. 4.2) (Zipser & Bennett 1976). The flow of
sensory information to the midbrain is briefly blanked out at the
moment when the fish expects to fire its own electric organ. Hence,
the fish is electrically deaf to itself and can only hear the discharges
of other fish on this sensory channel.

In the central nervous system, the neural projections of
Knollenorgans (see Fig. 4.2; Zipser & Bennett 1976). The flow of
by Bell (1979)) which are separate from the projections of
ampullary receptors and mormyromasts. Lesions to one midbrain
nucleus, known as the nucleus mesencephali exterolateralis,
which receives only Knollenorgan input, abolish a communication
response in which the fish briefly goes silent upon 'hearing' the
electric discharge of another (Moller & Szabo 1981). The lesions
do not affect the fish's ability to sense objects by electrolocation.

Mormyrids have segregated the communicatory and non-
communicatory sensory functions for electroreceptors in a way
that is remarkable for vertebrates. The sensitive Knollenorgan
receptors, functioning solely in communication, behave as filters
matched to the species-specific communication signal. In the
central nervous system, there is neuronal circuitry to eliminate
jamming from the fish's own signals—yet another adaptation to
the specific function of communication sensing.

4.2.4   Visual chasing in flies

A final example of a communication sensor is taken from visually
guided mating behaviour in dipteran flies. As in two of the

previous examples, the sensor of interest is sexually dimorphic (Fig. 4.3), as is the behaviour. Male flies pursue moving objects in a high-speed, carefully manoeuvred chase. Chases appear to function in reproduction for, when a male chases a female, he often grabs her in a sudden 'marriage by capture' (Land & Collett 1974; Wehrhaln 1979; review by Wehner 1981). Females do not chase as readily as males (Land & Collett 1974), although female–male chases do occur occasionally. Chasing males tend to keep the target in the upper-front part of their visual field, whereas females use the lower-front field; males tend to control their velocity so as to maintain a constant distance from the target, while the female's velocity is poorly regulated as the target twists and turns in space (Wehrhaln 1979). Chase paths and the probable control mechanisms of chases were analysed by Land and Collett (1974).

The visual systems of these flies and many other Diptera are sexually dimorphic. The compound eye of the fly is made up of 2500–3200 ommatidia, slightly fewer in females than in males. Males possess a central patch of large-facetted ommatidia on each eye, making a central fovea. Females of many species lack this central fovea (see example from *Syritta* in Fig. 4.3). The large facets improve the spatial resolving power of the central ommatidia because of reduced diffraction. In *Musca* the sexual dimorphism is more subtle. Each ommatidium is composed of eight photoreceptors, which are arranged as two centrally located cells (R7 and R8) surrounded by six larger cells (R1–R6). In females, the absorption pigments of the central photoreceptors within an ommatidium are usually different from those of the surrounding six cells. In males, the pigment in R7 and R8 is the same as in the surrounding cells in the dorsal central fovea of the eye (Franceschini *et al.* 1981). The male seems to have sacrificed colour vision in the fovea for the sake of increased light-capturing ability. Such a sacrifice may be crucial if he is to follow a small visual target at high velocity. The specialised retina of these flies projects to a neuropil region of the fly eye known as the lamina. Here too, there are sexually dimorphic interneurons, at least three of which are giant interneurons found in males but not in females (Hausen & Strausfield 1980).

The argument that the dorsal fovea in Diptera is playing a special role in communication is based on the sex difference in the structure of the sense organ. Males pursue females during sexual behaviour, but not vice versa. Because mating depends on this chase, the receptor has evolved so as to enhance chasing abilities, hence the sexual dimorphism.

The above four examples were selected because they illustrate how receptors can become refined for reception of communication signals, and they are, to some extent, exceptional. More typically, we think of the evolution of a sensory modality being under the influence of many types of selection pressure, not just those derived from communication. However, as the sections below will illustrate, these sensory receptors also function in the context of communication, and are often adapted specifically for receiving communication signals.

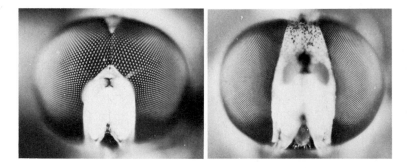

Fig. 4.3. The dorsal fovea in the eye of the dipteran *Syritta pipiens* is sexually dimorphic. The male (left) has a concentration of large ommatidia near the dorsal mid-line which the female (right) lacks (scale = $100 \, \mu$m). (From Land 1981.) In other Diptera, like *Fannia* and *Musca*, which are less dimorphic in the eyes, males appear to be far better than females at high-speed chases of moving targets. Males must chase and capture females in order to mate with them. They keep the female in their dorsal frontal field of view, using this specialised fovea. Females, if they chase at all, use the lower part of the eye.

For some sensory systems, the view of the world, like that through a small window, is only a small fraction of the total picture. The view through this window will be more or less suited to the behavioural needs of the animal, and since communication signals must pass through this window, we should learn something of its properties.

4.3 Sensory filtering

A communications engineer puts considerable thought into the design of filters for a receiver. Weighing the advantages of a broad filter which responds to a wide range of signal types against the advantages of a narrow filter which cuts out noise in the communication channel, the engineer must specify the limits of filter sensitivity (the bandwidth), the steepness of the filter's cut-off, whether to have a single type of filter or to use many different filters in parallel, and other related characteristics. How are these decisions made? How are filters designed to meet the desired specifications? Two different engineering methods are discussed here, frequency-based filtering (section 4.3.1), and matched filtering (section 4.3.2), and each is related to sensory filtering in animal sensory systems.

4.3.1 Frequency-based filtering

Frequency-based filtering relies on the fact that time-varying signals can be split into their different component frequencies. Mathematically, this is described as Fourier analysis (see Bracewell 1978), which uses the fact that any time-varying function may be expressed as a sum of sinusoids of varying amplitude, frequency, and phase lag (delay). Most physical stimuli used in animal communication can be viewed as time-varying signals suitable for Fourier analysis. For example, sound pressure changes set up by a source of sound, electric currents generated by electric fish, and the amplitude of optical signals may all vary with respect to time. Visual signals may also be considered as electromagnetic radiation of extremely high frequency. Chemical signals are not conveniently analysed using Fourier techniques because the time-varying component of a chemical signal is probably of minor importance compared to the structure of the chemical.

The sound spectrograph uses frequency-based analysis to draw a picture of sound (for voice, this is sometimes called a voiceprint or visible speech—see Cherry 1966). The amplitude coefficients (but not phase angles) for each component frequency are displayed for short time segments of the signal strung together to paint the picture (sonagram) of the sound (as in Fig. 4.1a). Frequency or Fourier techniques may be applied to signals which vary in space rather than in time; in this case, however, the analysis is usually conducted in two space dimensions (horizontal and vertical, for example) rather than in the one time dimension. A visual scene may be represented as a two-dimensional map of spatial frequencies (cycles per visual angle rather than cycles per second).

Once a communication signal is represented in the frequency domain, then a filter can be designed which has maximum 'gain' (i.e. amplification) for frequencies where there is signal energy, but minimal gain where there is none (see Fig. 4.4). The output of the filter, which is the product of the spectrum of the signal with the gain of the filter, will enhance signal frequencies and suppress non-signal frequencies. The advantage of selecting the filter according to the spectrum of the desired signal, of course, is that unwanted signals or noise which have different spectral characteristics will be reduced relative to the signal, and hence the signal-to-noise ratio will be improved. To obtain the maximal signal energy compared to noise energy, the profile of the gain curve should match the profile of the power spectrum of the expected signal (see Fig. 4.4a).

Frequency-based filtering in audition

The calling song of the male cricket *Gryllus campestris* is composed of four syllables, one soft tone burst about 10–20 ms long followed by three louder tones peaking at 60 ms intervals. The Fourier analysis of the entire song yields a complex power spectrum in which the primary power peak occurs at 4 kHz and a lesser peak occurs at 16 kHz. When this is compared with the thresholds for spike initiation in single fibres in auditory nerves to sinusoidal

Fig. 4.4. Two filtering strategies for enhancing weak signals masked by noise in the communication channel. (a) In frequency-based filtering, the 'gain' curve for the receiver should follow the power spectrum of the expected signal (upper right). The spectral characteristics of the output signal are determined by taking the product of the spectrum of the input signal with the gain of the filter. (b) In matched filtering, the output of the filter is the cross-correlation between the received signal and an expected signal (a template). The cross-correlation function slides past the received signal looking for the best correlation. If the received signal differs significantly from the expected, as in 2, the correlation is near zero; if the received signal matches the expected, then the cross-correlation will be significant. A matched filter is an optimum linear filter for detecting signal in noise.

stimulation, the threshold curves follow the inverse of the call spectrum almost exactly (Huber 1977; see Fig. 4.5). Hill (1974), referring to the peak of the power spectrum of the individual sound bursts (chirps) of the call as the 'carrier frequency', in order to distinguish it from the repetition frequency of chirps, showed

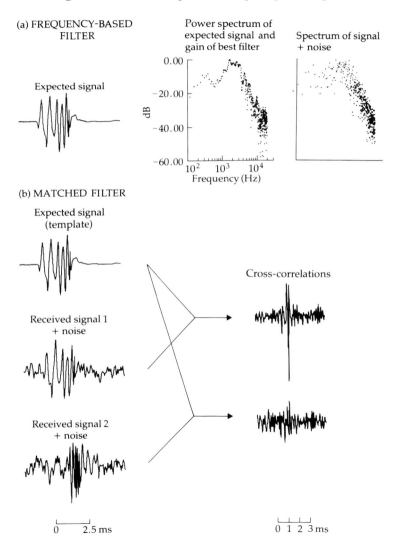

(a) FREQUENCY-BASED FILTER

Expected signal

Power spectrum of expected signal and gain of best filter

Spectrum of signal + noise

(b) MATCHED FILTER

Expected signal (template)

Cross-correlations

Received signal 1 + noise

Received signal 2 + noise

0   2.5 ms

0 1 2 3 ms

Chapter 4

that *Tellogryllus commodus* females approached tone pips with sine-wave frequencies matching the carrier frequency (about 4 kHz). Hill also demonstrated that acoustic units in the cervical connective were tuned to the same frequency, although the single-unit tuning curves were broader than the range of frequencies to which behavioural sensitivity could be demonstrated (the audiogram).

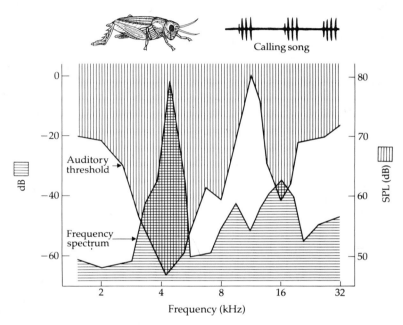

Fig. 4.5. Comparison of the auditory sensitivity of *Gryllus campestris* with the spectrum of the calling song of the male. The auditory threshold is measured in decibels SPL, the spectrum of the sound intensity is measured in decibels relative to an arbitrary standard. (From Huber 1977.)

The antenna of mosquitoes has a greatly enlarged Johnston's organ, which serves an auditory function. The male's antenna appears to resonate mechanically to sound frequencies near 400 Hz, and this serves to filter out all incoming sounds except the flight sounds of the female of the same species (Roth 1948). *Drosophila* males produce weak sounds by beating their wings during courtship (Bennet-Clark & Ewing 1967, 1969); the feathery aristae on the antenna are sensitised to sounds of 180–220 Hz (Bennet-Clark 1971) by a mechanical resonance. Fiddler crabs are sensitive to sounds transmitted through the substrate, and vibration sensors in the legs are tuned to the frequencies in the sounds produced by rapping the claw on the ground.

Some vertebrates exhibit similar matching between auditory sensitivity and a predominant communication signal. Behavioural audiograms are now available for a number of fishes (Popper & Fay 1973). For six species of damselfishes (*Eupomacentrus*) these are very broad; the sensitivity range is from 200 to 900 Hz, with a minimum threshold at 500 Hz (Myrberg & Spires 1980). *Eupomacentrus diencaeus* produces 'chirps' during courtship which have maximal energy between 450 and 900 Hz, accurately matching the audiogram. The ear of the toadfish, *Opsanus tau*, is similarly tuned to the courtship sounds for this species (Fish & Offutt 1972). Some fishes can hear frequencies of 200 Hz and below, and are probably sensing particle motion rather than sound pressure; others, such as the ostariophysi, appear to be sensitive to sound pressure (see discussion in Hawkins 1981).

In frogs and toads, the correspondence between the frequency sensitivity of single fibres in the auditory nerve and the species-specific mating call was noted by Capranica (1965; see also Frishkopf & Goldstein 1963; Frishkopf *et al.* 1968). The power spectrum of the mating call of the bullfrog *Rana catesbeiana* has two peaks, one at 250–300 Hz, and the other at 1600 Hz. The individual tuning curves of auditory nerve fibres are V-shaped, and come in three varieties, low frequency (200 to 400 Hz), mid frequency (600 to 800 Hz), and high frequency (1400 to 1600 Hz), similar to *Eleutherodactylus coqui* (section 4.2.2). Thus no one auditory receptor is a perfect match to the spectrum of the mating call. Rather, the combined range of sensitivity of the three receptor types matches the spectrum of the call of the male. The story for a number of other frog species is much the same: two or three classes of unit are tuned to different ranges of frequencies, but always with a close correspondence to the adult male mating call (see Capranica 1976; Capranica & Moffat 1975; Loftus-Hills & Johnstone 1970). One of the most interesting cases is that of the cricket frog (*Acris crepitans*), which shows geographical variation in its call structure, especially in the high-frequency components. The calls of frogs from New Jersey have a peak at about 3500 Hz, while those of animals from North Dakota peak at about 2900 Hz. This is associated with a shift in the tuning of auditory fibres projecting from the basilar papilla, from 3500 to 2900 Hz, corresponding to this regional variation in the call. There is only one class of unit in the low-frequency range, 200–1000 Hz, presumably projecting from the amphibian papilla, and there is no geographical variation in tuning within this frequency range (Capranica *et al.* 1973).

In birds and mammals, where communication sounds tend to be the most varied among vertebrates, frequency-tuning characteristics of individual auditory receptors (hair cells) are sharper than those of most amphibians and fishes, and there is more diversity of best-frequency types, because of the specialised mechanical properties of the cochlea. Rather than sensitivity to a small number of discrete frequencies, birds and mammals usually possess a continuous family of filters, each tuned to a narrow part of the spectrum of sensitivity (see Schwartzkopf 1949; Dooling *et al.* 1971; Konishi 1970). Nerve fibres tend to be ordered according to best frequency, as in the cochlea, a pattern which is termed 'tonotopy'. Konishi looked at single units in the auditory cochlear nucleus of song-birds and found many individual units with best frequencies lying in the range employed in bird song, but nothing like the frequency filtering specialisation seen in fishes and amphibians.

Frequency-based filtering in vision

For optical systems, frequency-based filtering usually refers to colour sensitivity. Aspects of spatial frequency filtering will not be considered here. In aquatic environments, where turbidity, water colour and depth result in low light levels and poor horizontal light penetration, there are unusually severe problems for the signal receiver (see section 3.3.3). Yet vision is one of the best-developed senses in fishes, and serves a primary function in detecting prey, predators and conspecifics. There is a striking correlation between the absorption spectrum of the rhodopsin pigment in the retina and the absorption properties of the water in which a fish is living. This is most convincingly demonstrated by comparing deep-sea environments, which have monochromatic deep blue colour, shallow-water coastal zones, which have a yellow to yellow–green cast because of the rich supply of phyto-plankton, and fresh water, which is coloured yellow–green to yellow or even red.

Lythgoe (1972, 1979) suggests that the absorption spectrum of the visual pigments in the fish's retina change according to the water colour for two adaptive reasons: to increase sensitivity and to increase visual contrast. If a grey object near to the surface of the water is viewed underwater horizontally and at close range, the spectral radiance from the target will be nearly grey, because the target is reflecting the white light shining on it from the sunlight above (d = 0 in Fig. 4.6a). When the object is moved away from the observer, however, the horizontal water column absorbs the reflected light selectively, resulting in a peak in the radiance spectrum at the transmission maximum (T_{max}) for the water. The background light against which the object is seen, a result of scattering, also has a peak radiance at T_{max}. The radiance spectrum for the object will match that for the backlight when the object is moved far away (d = infinity in Fig. 4.6), so the object will become invisible. At closer range, however, the object will appear darker than the backlight for wavelengths near to T_{max}, and will appear brighter than the backlight for wavelengths well above or below T_{max} (regions A and B in Fig. 4.6b). At two crossover wavelengths, λ_1 and λ_2, the object and backlight will have identical radiances. A plot of the visual contrast between the grey object and the backlight (Fig. 4.6b) shows that, for wavelengths near T_{max}, contrast declines rapidly as the object is moved away, but for wavelengths above and below these critical values, it remains high as the object moves away. Thus for optimal contrast, the visual pigment should have its peak absorption offset from T_{max}, either above or below.

In deep water the situation changes; now the object is il-luminated from above by light which has filtered down through the water column. The scattered background light, also filtered through the water column, has the same spectral radiance as the light reflected from the object. Since contrast in this case is not wavelength dependent, the best a fish can do is to enhance its absolute sensitivity to light by having a pigment matched to T_{max}. Lythgoe and others have collected specimens of fish in deep marine environments, shallow coastal zones, estuaries and fresh water, and confirm the prediction that pigments should be

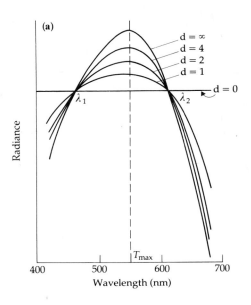

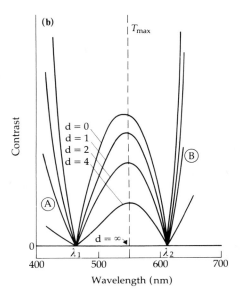

matched to T_{max} in deep water and offset in shallow water (see Lythgoe 1979 for references, and section 3.3.3).

Frequency-based filtering in electroreception

Electroreceptors in fishes behave as band-pass filters of electric stimuli. Gymnotoid fishes have a type of electroreceptor classed as tuberous. In wave-discharging species, these electroreceptors are tuned to the fundamental frequency of the electric discharge of the species or, more precisely, to the frequency of the individual's own discharge. Figure 4.7 shows three species of wave-discharging fish from South America which have distinct ranges of discharge frequencies, and non-overlapping tuning curves to match. Tuning in electroreceptors, unlike the mechanically based tuning of auditory receptors, is based on the electrical oscillatory properties of the receptor cell, which is a modified hair cell (Hopkins 1976).

Gymnotoids which produce brief pulses rather than continuous tones also have tuned electroreceptors; this tuning is sensitive to the peak of the power spectrum of a single pulse, i.e. the carrier frequency (Hopkins & Heiligenberg 1978; Bastian 1976). A similar situation holds for the African mormyrid fishes, which also produce pulse discharges. The high-frequency communication sensor known as the Knollenorgan, already discussed in section 4.2.3, is also tuned to match the peak of the power spectrum of a single pulse typical for the species (Hopkins 1981).

4.3.2   *Matched filters and communication*

A second type of filter, recognised by communication engineers as superior to the frequency-based filter in its ability to detect signal

---

Fig. 4.6. (a) The spectral radiance from a neutral grey target suspended in water near the surface at distances of 0, 1, 2, and 4 metres, and at very large distances. At very close range, the radiance curve is flat white. The water has a spectral transmission maximum at the wavelength T_{max}, indicated by a dashed line. At long viewing distances (d = infinity), the object is indistinguishable from the veil of the background spacelight. (b) Visual contrast plotted as a function of wavelength for the same situation as in (a). At d = infinity, there is zero contrast between spacelight and light reflected from the object, so the object is invisible. At reduced distances, contrast varies with wavelength. Contrast is best at wavelengths well below and above T_{max} (zones A and B). To achieve maximal contrast under these conditions, visual pigments should be offset from the transmission maximum of the water.

in the presence of noise, is called a 'matched filter', a 'correlation filter' or, simply, an 'optimal filter'. Matched filters operate in the time domain, not in the frequency domain, and perform a cross-correlation between the temporal pattern of the received signal

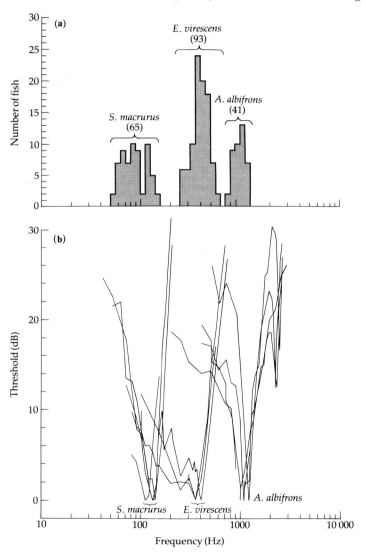

and a 'template' of the desired signal (see Fig. 4.4b). The matched filter may be used in some biological sensors, but the concept is relatively new and few biological systems have been adequately analysed.

Lee (1960) demonstrates mathematically that a cross-correlation detector is an optimal linear filter for detecting a known or expected signal in the presence of noise. A cross-correlator is

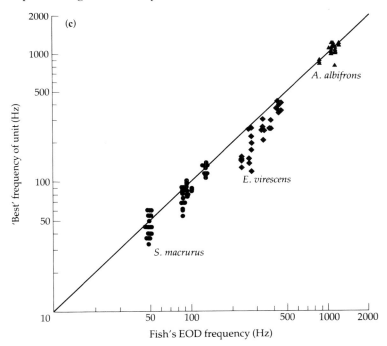

Fig. 4.7. Illustration of the close correspondence between the electric organ discharge frequency of wave-discharging gymnotoids and tuning of electroreceptors. Three sympatric species of gymnotoids are shown: *Sternopygus macrurus, Eigenmannia virescens,* and *Apteronotus albifrons.* **(a)** Histogram of fundamental frequencies of discharges of individual fish. All frequencies are measured at 25°C. The frequencies are species-specific and show no overlap. **(b)** Tuning curves for tuberous electroreceptors. The ordinate is the threshold of the unit relative to the most sensitive value, the abscissa is the frequency of the sine wave used as stimulus. There is a close correspondence between a unit's best frequency and the discharge frequency for the species. **(c)** An individual's discharge frequency is matched by the tuning characteristics of its own electroreceptors. (Modified from Hopkins 1976.)

essentially a template signal which slides past a received signal (in time) looking for a perfect match or correlation (see Fig. 4.4b; Bracewell 1978 provides a more formal discussion).

A matched filter specifies a *single* desired stimulus. All other signals will be received less efficiently, thereby severely limiting the repertoire of signals correctly received in the communication channel. However, one can imagine a parallel series of matched filters, each designed to receive one specified signal. Alternatively, a single matched filter would be highly effective in cases where messages are encoded as modulations of a carrier signal. The receiver, in this case, is matched to the carrier, but still receives a variety of messages because of the variety of modulations of the carrier.

Two examples of matched filters will be given: first, that for detecting the complex frequency-modulated echolocation cries of a bat, and secondly, that for detecting the carrier pulse of an electric discharge. In contrast to the previous examples, the examples of matched filters presume involvement of the central nervous system, not just the peripheral receptors.

Matched filtering in echolocating bats

Echolocation is not strictly communication, because sender and receiver are the same individual. Nevertheless, bat echolocation is a good illustration of the matched-filter concept. In fact, engineers working in radar and sonar theory have shown that a matched-filter receiver is the best for these systems (see Woodward 1964); therefore, if bats have evolved optimal receivers, they too may use matched filters.

Evidence for matched filtering comes from a series of behavioural studies testing the accuracy of distance measurement by echolocating bats. Simmons (1973) trained bats to discriminate between a distant target and a near target in complete darkness; the only possible cue available coming from echoes bounced off the targets. The targets were brought closer and closer together to determine at what point confusion arose about the distance to the objects. Estimation of the distance of an object requires accurate measurement of the time of arrival of a signal; this must be very precise because, with sound travelling at 330 ms^{-1}, an error of one millisecond corresponds to a 17 cm error in target range. With most sounds, which have rather slow onsets and terminations, the exact time of arrival of the signal is sometimes ambiguous, leading to uncertainty in the position of the target in space. Furthermore, if the target is moving towards or away from the bat, the echo will be shifted up or down in frequency due to the Doppler effect, further complicating the estimation of the arrival time. The best that the bat can do is to use the cross-correlation between the emitted sound and the received echo, the time of the peak of the cross-correlation being the best estimate of the echo delay (Woodward 1964). Simmons' conditioning experiments demonstrate that many bats which produce frequency-modulated cries (FM-bats) can do as well at echo-ranging as one would predict on the basis of the cross-correlation method. To obtain an estimate of the best possible performance, Simmons used the signal auto-correlation (correlation with itself), because he assumed that an undistorted echo (with no distortion and no noise) would give the best result. Plots of errors in target range discrimination as a function of range difference fit the envelope of the auto-correlation function perfectly. It appears, therefore, that bats are optimal receivers for their own signals. They may be performing a type of cross-correlation between their own sounds and the echoes of these, but this has not been shown physiologically.

Matched filtering in electroreception

A second example of a proposed matched filter is provided by electroreception in the pulse-discharging gymnotoid fish *Hypopomus* and the pulse-discharging mormyrid *Brienomyrus brachyistius*. The spectral sensitivity (tuning curve) of the electroreceptors in each case matches the power spectrum of the species-specific electric pulse. This is to be expected both from a matched filter and from a frequency-based filter. However, Heiligenberg and Altes (1978) and Hopkins and Bass (1981) altered the waveforms of some electrical stimuli without altering the power spectrum, and showed that the fish responded differently to them. Best responses were obtained for natural electric pulses. These fish, therefore, appear to be sensitive to events occurring in time and not just to frequency differences.

4.3.3   Conclusions on filtering

Peripheral sensory receptors which react to portions of the available energy spectrum perform only one aspect of the stimulus filtering which occurs in the nervous system. Integration between neurons accounts for the bulk of filtering, especially for complex features of stimuli. Peripheral filters, however, define an organism's windows to the world of stimuli. If stimuli are rejected by a peripheral filter, the animal has no hope of recovering this aspect of the stimulus through neural integration.

The match between peripheral filters in the auditory, electric, chemical and visual systems and the types of signals used in communication is perhaps surprising. Filtering carries advantages and disadvantages: whereas it improves the signal-to-noise ratio for those signals it is desired to pick up, filtering results in the loss of sensitivity to other stimuli. Engineering models for design of communication receivers are useful for characterising the adaptive significance of filtering for purposes of animal communication.

4.4   Neural coding and animal communication

Once a sense organ has filtered and transduced an incoming stimulus, the next step is the production of some pattern of nerve spikes which can travel within the nervous system to influence adjacent neurons and integrate responses. The translation of stimuli into nerve impulses is called *neural coding*. It takes place not just at sense organs, but also at each synapse in the nervous system.

4.4.1   Codes for stimulus amplitude

Most sense organs code for the amplitude of the stimulus; in other words, they signal, by some pattern of nerve activity, how much stimulus was present. The most common neural code for amplitude is the frequency code. Typically, over some range of stimulus intensity the frequency of nerve spikes is proportional to the stimulus amplitude. Below some critical intensity, the threshold, the firing frequency is zero, and at some upper limit of stimulus intensity the firing frequency reaches a plateau. Figure 4.8 illustrates frequency coding by an olfactory sensillum in *Bombyx*. Presented with odours at concentrations varying over four log units, the nerve fibre from the *sensillum trichodeum* responds with steadily increasing spike frequency.

Several input–output relationships have been described for frequency coding. In a linear relationship, spike frequency is linearly proportional to the stimulus intensity; in a logarithmic relationship spike frequency is proportional to the logarithm of the stimulus intensity; and in a power relationship the spike frequency is proportional to the stimulus intensity raised to a constant exponent or power. It is not known in what ways different forms of coding are adaptive for different sense organs.

Frequency codes are but one way of translating intensity into a pattern of nerve spikes. Perkel and Bullock (1968; see also Bullock 1981) proposed a number of other codes based upon variation of the microstructure of nerve spikes without changes in the mean spike rate.

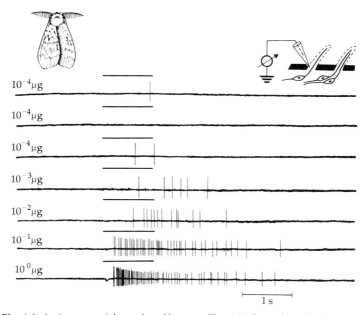

Fig. 4.8. Action potentials produced by a *sensillum tricodeum* of a male silkworm moth, *Bombyx mori,* in response to pheromone delivered at different concentrations. The receptor codes the stimulus intensity using a frequency code. (From Kaissling & Priesner 1970.)

In those electric fishes with pulse-like electric organ discharges, these EODs evoke responses in electroreceptors which code for amplitude in a number of ways. Mormyromast receptors in mormyrid electric fishes, and one class of tuberous electroreceptors in gymnotoid fishes, produce a volley of spikes the duration of which depends upon the EOD amplitude. These receptors are called burst-duration coders. Gymnotoid fishes with wave-like EODs have both 'T' receptors, which code intensity as the latency

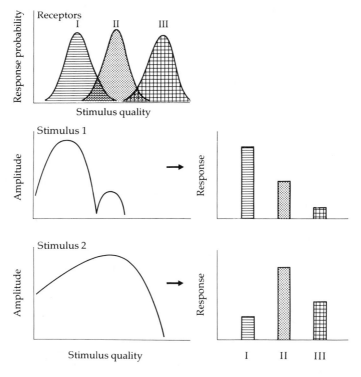

Fig. 4.9. Coding stimulus quality within a sensory modality by parallel processing. The upper left diagram shows the response probability of three types of receptors as a function of stimulus quality (stimulus quality refers to any continuously variable stimulus dimension such as frequency or wavelength). Below this, on the same abscissa, is the power spectrum of two different stimuli. To the right are the response probabilities of the three receptor classes to the two stimuli. To assess stimulus quality, it is necessary to compare responses across all three receptor types in parallel.

of the response compared to the stimulus cycle (so-called phase coders), and 'P' receptors, which code intensity as the probability of firing a spike on a given stimulus wave.

4.4.2   Codes for stimulus quality

Over a century ago, Johannes Müller outlined the concept of specific nerve energies, a theory that individual sensory nerves always give rise to the same characteristic sensation no matter how stimulated (thus we 'see' stars when struck on the eye). The modern extension of this notion is the concept of *labelled lines*, which applies not just to one sense modality compared with another, but to specific receptor types within a sensory modality.

The labelled-line concept, combined with variety in sensory filters, permits the neural coding of stimulus quality. To illustrate this, consider a hypothetical sensory system, as shown in Fig. 4.9. Three receptor classes are illustrated for this sensory modality, and each responds to a certain region of the quality spectrum (the abscissa), as indicated by the probability of firing (the ordinate). One could substitute 'sound frequency' for 'stimulus quality' when considering an auditory unit, or 'wavelength of light' as the equivalent in the visual system. Although the three receptor types show wide overlap in their tuning characteristics, they differ sufficiently that when presented with two stimuli which differ in their spectral characteristics, the three receptor populations respond differently (see lower part of Fig. 4.9). To the first stimulus, receptor 1 responds vigorously, receptor 2 less so and receptor 3 very little. For stimulus 2, receptor 2 gives the strongest response, while receptors 1 and 3 are weaker. Stimulus quality can thus be coded as the relative response intensity of a number of labelled lines acting in parallel. To interpret the code, the brain must be able to compare the strength of responses in all three receptor populations. Erickson (1974) discusses parallel processing further. The concept is clearly useful when applied to assessment of colour by the visual system. Visual pigments with broadly overlapping absorption spectra resemble the curves in Fig. 4.9. Erickson applies the concept to chemical stimuli; although the dimensions for stimulus quality are unknown in this modality, there appear to be neurons which have broad, but differing stimulus specificities.

Some sensory systems may translate different stimulus qualities into unique temporal patterns of nerve impulses rather than depending upon labelled lines. This may be especially important in the auditory system, although the mammalian cochlea is a fine example of a system of labelled lines, with individual nerve fibres signalling the presence of a narrow range of frequencies. However, low-frequency sounds elicit auditory nerve spikes which are phase-locked to the stimulus; in other words, the spikes are triggered cycle-by-cycle by the sound stimulus. The effect of phase locking is that the fundamental frequency of the stimulus is transmitted to the central nervous system as a temporal code, rather than as signals in a unique nerve fibre originating somewhere in the cochlea. Fay (1981) suggests that temporal coding may be the basis of sound frequency discrimination in fishes (which lack a cochlea). Hopkins and Bass (1981) conclude that temporal coding applies to wave-form discrimination in electric communication by fishes.

Neural coding, like any translation of a signal from one form to another, is a further component of stimulus filtering. All the properties of a communication sensor, its sensitivity, selectivity, gain, and vulnerability to masking and to confusion by similar signals, will be dependent upon mechanisms of coding, as they are upon mechanisms of filtering.

4.5   Recognition mechanisms in communication

Respondents select their responses to communication signals from a number of possibilities. That one is chosen rather than another is the ethologist's best evidence that the signals are recognised by the respondents (see section 1.3). There has been much speculation about the mechanisms of recognition of complex stimuli since Lorenz (1935) introduced the concept of an Innate Releasing Mechanism (IRM). The idea of the IRM as a unitary lock and key mechanism seems implausible and has been replaced by the more realistic view that recognition involves a series of stimulus filters in the nervous system. Peripheral receptor filtering and neural coding are obvious components of stimulus filtering, and must be incorporated into the recognition mechanisms. Similar to the IRM concept is the physiological concept of feature-detecting neurons, which are nerve cells, or groups of nerve cells, responding to characteristic features of a given stimulus. The famous 'bug-detector' neuron in the frog retina (Lettvin *et al.* 1959), and oriented-bar-detecting neurons in the visual cortex (Hubel & Wiesel 1959) illustrate the idea of feature detectors in sensory systems. Schmitt and Worden (1974) discuss aspects of feature detection from a physiological viewpoint.

4.5.1 *Mating-call detectors in frogs*

Capranica (1965) was one of the first to identify the cues essential to eliciting an adaptive response to auditory stimuli. He used the evoked vocal response of the bullfrog (*Rana catesbeiana*) and tested a variety of synthetic sounds. Two components were necessary to evoke calling: a low-frequency component at about 200 Hz, and a high-frequency component at about 1600 Hz. Presented separately, these components were ineffective, but together they evoked calling. Addition of sound in the intermediate frequency range, around 500 Hz, inhibited calling.

To explain their behavioural observations, Capranica and his colleagues have explored response characteristics of units throughout the auditory system in frogs, from the peripheral nerve to the dorsal medullary nucleus, superior olive, torus semi-circularis and auditory thalamus (see Capranica 1976). The identification of peripheral units tuned to low, mid and high frequencies has already been discussed in section 4.2.2 for another species of frog, and results were similar for the bullfrog. It is only in the auditory thalamus that Mudry *et al.* (1977) found evidence for any comparison of the inputs on these separate frequency channels. There, in contrast to results with more peripheral units, evoked potentials indicate that low and high frequencies must both be present to elicit responses; when they are presented separately, the evoked potential is much reduced. Units in the thalamus thus appear to behave like logical AND gates, which fire only when both low- *and* high-frequency stimuli are present.

4.5.2 *Complex signal recognition*

Although there are extensive descriptions of bird songs, there is as yet little information about the acoustic parameters which are relevant to species recognition. Some notable exceptions come from the work of Emlen (1972; see section 2.3), who broke down indigo bunting (*Passerina cyanea*) song into its component syllables, scrambled them and played them back to territorial males, and of Falls (1963), who created artificial white-throated sparrow (*Zonotrichia albicollis*) song using pure tones. Recently, with the opportunity of using computers to generate an endless variety of synthetic songs, there has been new interest in defining key stimulus variables evoking recognition responses in audition.

Peters *et al.* (1980) have shown in song sparrows (*Melospiza melodia*) and swamp sparrows (*M. georgiana*) that the temporal order of syllables and their repetition rate is less important for recognition by males of territory owners, than is the type of syllable used. Marler and Peters (1977) also showed a preference in young birds to learn songs if the syllables were of the species-specific types, even when the tempo of the songs had been rearranged.

Efforts to understand auditory processing beyond the auditory nerve, and to apply the results to species-specific sound, are just beginning with birds (Bonke *et al.* 1979; Scheich *et al.* 1979; Leppelsack & Vogt 1976). While units which respond vigorously to species-specific calls can be found in the forebrain of birds, it is not yet clear how to characterise the specificity of the unit. What are the appropriate dimensions to vary in order to specify the unit's tuning?

For mammals, the best evidence for specialisation of auditory units to species-specific sounds comes from bats, where both the peripheral and central auditory systems are specialised for receiving the ultrasonic cries used in echolocation. Since this is not communication in the strict sense, it will not be discussed here, but is considered further by Suga and O'Neill (1979) and by Neuweiler (1979). Analysis of the sensory processing of complex stimuli will undoubtedly be a fruitful area for future research. Ewert *et al.* (in press) give examples from studies of recognition of bird song, face-detecting neurons in the monkey cortex, visual pattern recognition in newts and in cichlid fishes, and processing of the jamming avoidance response in electric fish.

4.6   Signal localisation

Up to this point, we have considered peripheral and central sensory mechanisms that permit receivers first to detect and then to discriminate between stimuli. Now we turn to the problem of locating the signal in space.

4.6.1 *General problems in signal localisation*

Localisation of a signal and its recognition require very different sensory hardware, yet spatial cues carry a great deal of information which may be of value to the receiver. In some cases, such cues may indicate identity of the signaller (see section 2.5). This was demonstrated nicely for territorial song-birds by Falls and Brooks (1975), who showed that a territory-owning white-throated sparrow would show little response to its neighbour's song played to it from the appropriate territory, but would react aggressively when the song came from a new, unexpected location. Often, when an animal responds to a signal it assumes some orientation in relation to that signal. Signal localisation, therefore, is an integral part of communication, but the mechanisms involved differ widely for each sensory modality.

To locate a signal, the receiver needs to detect some vector quality of the physical signal which is indicative of direction. Three signal vectors are especially useful:

(1) the gradient of signal amplitude, that is, the change in the amplitude of the signal with distance and direction;

(2) the directional vector of the signal energy itself, such as the particle motion in a sound signal; and

(3) the signal velocity, an inherently directional characteristic which indicates both speed of signal travel (a scalar) and its direction (a vector).

Gradient detection works best where the gradient is steep, in other words, where the change in amplitude is rapid with respect to the animal's body size. Velocity detection is favoured only when the signal speed is slow enough to measure time differences in signal arrival; detecting energy vectors indicative of direction only works if the animal possesses a directional receiver, like the polarised hair cells of the lateral line system.

An amplitude gradient may be detected either by making instantaneous amplitude measurements at two or more locations separated in space on the animal's body, or by making successive measurements of amplitude at two different places using a single receptor organ. This distinction applies not just to communication, but also to animal orientation. Fraenkel and Gunn (1940) referred to orientation to a stimulus gradient by successive amplitude measurements as *klinotaxis,* to distinguish it from various other taxes which involve instantaneous measurements. Gradient detection by successive measurements is a necessity for the oriented swimming behaviour of bacteria which are known to migrate toward attractant substances in the fluid medium, and away from repellent substances. The size of an individual bacterium is too small to permit the detection of a gradient of chemical concentration across the cell surface. Bacteria such as *E. coli* must therefore make successive measurements of chemical concentrations and briefly 'remember' them. The frequency of their tumbling (turning randomly in space) is low as long as the cell is moving towards an increasing concentration of attractant, but is high when the cell is moving towards a decreasing concentration of attractant (see review by Adler 1976). Insects probably follow gradients of sex-attractant pheromone by similar comparisons because, owing to the long distances involved, the gradients will be shallow with respect to their body size and hence the separation of their receptor organs. Gradients would be steeper, according to theoretical predictions by Bossert (1968), if the signaller were to pulse its chemical emissions (see section 3.2.2). For long-distance sex attraction in moths like *Bombyx,* orientation probably depends instead upon detection of the direction of the wind carrying the odour molecules.

Some insects appear to be able to detect an odour gradient by comparing the simultaneous input through the right and left antennae. Under experimental conditions, Martin (1964) showed that honeybees could learn to turn toward the more concentrated of two odour sources presented in a Y-maze. He then placed tiny glass tubes with odours trapped inside over each antenna and found that bees could make instantaneous comparisons of the odour concentrations as long as these differed by a factor of at least 2.5 (ratio of 1:2.5). Whether bees detect natural odour gradients is unknown. Ants, which follow odour trails laid on the substrate, are close enough to the odour source to detect its steep gradient; if the antennae are glued so that they are crossed over, the searches of an ant oscillate wildly about the odour trail. Chemical orientation mechanisms, from single cells to vertebrates, are reviewed by Bell and Tobin (1982).

Physical stimuli which can be described as travelling waves, like light or sound, may be shaded, reflected and focused by accessory sensory structures. By these means, receptors may gain directional selectivity. This works especially well when the wavelength of the stimulus is short, because images and shadows will then tend to be sharp rather than blurred by diffraction (see section 3.4.3).

The pinnae of many vertebrate ears appear to reflect and shade sounds in order to impart directionality to acoustic stimuli (Knudsen & Konishi 1979; Grinnell & Schnitzler 1980; Shaw 1974). The highly complex movements of the pinna in cats, ungulates, bats and other mammals undoubtedly aid sound localisation.

Photoreceptors are made directional by the action of lenses which concentrate the light from one direction, and by pigment barriers between individual photoreceptors. The optical properties of visual systems vary enormously; some species, such as humans, are able to resolve up to 4000 lines per radian, while arthropods, with compound eyes, resolve only 5 to 30 lines per radian (see Land 1981 and Kirschfeld 1976 for reviews). Electroreceptors are directional only to the extent that they respond to the component of electric current which is perpendicular to the surface of the skin (Heiligenberg 1977).

For vision, electroreception and the tactile senses, directional or topographically separate receptors tend to form ordered surfaces, or maps, on to which stimuli from different directions project with different patterns of excitation. These surfaces or maps then contain spatial information which is conveyed to the central nervous system in an ordered way. The retina of vertebrates and the compound eye of arthropods are maps of visual signal intensities and colours in space; electroreceptors are laid out in a map configuration on the surface of the skin of electric fish, as are the tactile sense organs, such as the lateral line. Maps of sensory arrays may be the most efficient neural geometry for complex comparisons between neighbouring receptors to occur, as in, for example, lateral inhibition.

4.6.2   *The special case of sound localisation*

Because of their long wavelengths, sounds are difficult to locate in space. Three intriguing biological solutions will be discussed here:

how crickets use sounds to find their mates; how owls locate sounds and thus find their prey; and how fish locate sounds underwater.

How crickets find their mates

Many insects which stridulate, producing sounds by rubbing specially adapted parts of their legs and wings together, perform accurate 'phonotaxis' when presented with these species-specific sounds (Popov & Shuvalov 1977; Murphy & Zaretsky 1972). Yet most insects' bodies are too small to cast a significant sound shadow, or to focus or reflect sounds with frequencies below 10 kHz. This is because the diffraction of sound by objects is wavelength dependent, and sounds of 10 kHz, which have wavelengths of 33 mm, will merely pass unaffected around an object like a cricket. Two models have been proposed to explain how the cricket can extract directional information from sounds despite these problems. The first assumes that the ear is a simple pressure transducer, much like the ears of mammals, and that sounds are located by comparing the input to the two ears. The second model assumes that the ear responds to the gradient of sound pressure. The majority of evidence favours the pressure-gradient receiver model for crickets, although there is still some debate as to the details.

A sound-pressure receiver consists of a closed chamber with a thin, vibrating membrane exposed to incident sound. Local changes in sound pressure cause motion in the membrane, and this is detected mechanically. Since sound pressure is a scalar quantity, not a vector, the detector is not inherently directional; instead, it is made directional by external structures, such as the pinna, or by the presence of the head.

By contrast, a pressure-gradient receiver has two ports open to the incident sound, which can thus influence both sides of a vibrating membrane. When the sound pressure is equal on the two sides of the membrane, as would occur whenever the two ports lie at equal distances from the sound source along the expanding wavefront, the effects from the two sides cancel and the membrane does not move. When the axis between the two ports is turned perpendicular to the wavefront, then the pressure difference is greatest, and the membrane moves maximally (see also section 3.4.3).

The cricket's ears are located at the proximal ends of the tibiae of the forelegs. Autrum (1940) first recognised that pressure-gradient reception might occur in gryllids and bush crickets, which have a pair of tympanic membranes for each ear. A large tracheal tube which divides into posterior and anterior chambers at the level of the ear reaches the back of both these membranes. The tympana are also exposed to the outside air and incident sound and are connected to each other via the internal tracheal pathway. Autrum proposed that the sound incident upon the tympanic membranes would first act upon the outside and then penetrate through to the interior face of the opposite membrane. This creates a fine-tuned mechanism for directional reception. Although the cricket should be deaf to sounds in certain directions, this deficit is presumably compensated for by the other ear, as this is oriented differently.

Some crickets have only one tympanic membrane, but a tracheal tube leading to the other ear or to spiracles on the body surface may provide sufficient sound input to the back of the tympanum for a slightly modified pressure-gradient system to operate (see Fig. 4.10). To see if this works, one must determine not only the angle between the second ear and the first with respect to the incident sound, but also the path length within the trachea. The gain of the tracheal tube will also determine the extent to which the signals cancel out. These complexities are discussed by Lewis (1974), Hill and Boyan (1976), Nocke (1975) and Michelsen and Larsen (1978).

Through the use of free field sound stimulation, electrophysiological recordings have shown that a single cricket ear is highly directional without the need for neural interaction with the ear on the opposite side (Rheinlaender & Römer 1980; Nocke 1975). Figure 4.10 shows the influence of sound direction on spike activity of acoustic interneurons of an Australian bush cricket, *Tettigonia viridissima*. The response activity always falls for sounds from the other side of the body. Note that the sharpest rate of change of response occurs near the mid-line. Thus, when sounds come from near the mid-line, both ears will be stimulated equally. Slight deviations from the mid-line will cause large changes in the *difference* in left and right responses. The cricket should be most accurate in locating sounds which are straight ahead. Significantly, as females approach species-typical sounds in an open field, they take a zig-zag path (Huber 1977).

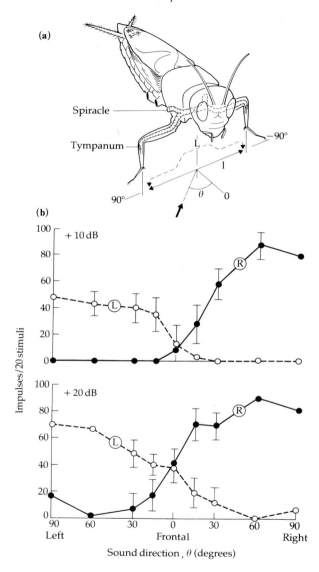

Fig. 4.10. (a) Drawing of the cricket *Teleogryllus commodus* showing the two pathways for sound to affect the tibial auditory organ. Sounds reach the interior of the tympanic membrane through the tympanum; they reach the interior of the same membrane by acting on the opposite ear, and transmitting through the

Vertebrate sound localisation

Pressure-gradient detection mechanisms may be important in some vertebrates. The problem encountered by insects will apply here also because the commonest signal frequencies used (1–6 kHz; wavelengths 340–50 mm) will be only weakly diffracted from the small heads (1–4 cm in diameter) of most vertebrates such as frogs, birds and small mammals (see Lewis & Coles 1980). Measurements in quail of cochlear microphonic potentials, which indicate the extent to which the ear has been stimulated by a sound, showed that each ear has a directionality at 315 Hz like that expected for a pressure-gradient detector (i.e. kidney-shaped on polar graph paper). The two ears are connected by an air-filled passageway through the skull, and this may provide the two inputs necessary for a pressure-gradient system to operate.

Owls are perhaps the most specialised and accurate of land vertebrates at locating sounds. They use their abilities in prey detection rather than in communication, but nevertheless give a fine indication of what can be achieved. The barn own (*Tyto alba*) can locate sound to within 2–3° in the horizontal plane and to within 5–15° in the vertical.

Two distinctive features of the owl's external auditory system appear to contribute to this accuracy. First, the right and left external ear canals open asymmetrically on the face, with the left higher than the right. Secondly, a dense ring of feathers makes up the facial ruff and these serve as a parabolic reflector of sound. Payne (1971) mounted small microphones in the ear canals of a stuffed owl and showed that the left ear is most sensitive to sounds

---

air-filled trachea. The net effect on the membrane is the *difference* in sound pressures from these two sources. When sound is incident at an angle θ to the body axis, the pathway to the interior of the ipsilateral membrane via the contralateral ear exceeds the path to the exterior by a distance of $l\sin\theta \times L$ (where l is the distance between the ears and L is the length of the tracheal pathway). The sound pressures at the exterior and interior thus differ in phase by $2\pi(l\sin\theta + L)/\lambda$, where λ is the wavelength of the sound. (From Hill & Boyan 1976.) **(b)** The directional characteristics of auditory interneurons in the left (O---O---O) and right ●——●——● sides of the bush cricket *Tettigonia viridissima* for stimulus frequencies of 20 kHz and two different sound intensities (10 and 20 dB). Inputs from the right and left sides are matched when sounds come from directly ahead. When the sound deviates from the mid-line, the two sides become mismatched and one side fires more vigorously than the other. (From Rheinlaender & Römer 1980.)

on the left side coming from below, and the right ear is most sensitive to sounds on the right coming from above. Knudsen and Konishi (1979) found that attenuating the sound entering the left side with loose cotton caused the owl to err to the right and too high. Removal of the facial ruff caused errors, but only in elevation, not in the horizontal plane. These experiments show the importance of binaural measurements of sound pressure for sound localisation.

One striking finding about the owl's auditory system is that higher-order auditory neurons can be found which respond to sounds in only a limited region of space, and these neurons are organised into a coherent map in the midbrain. Nowhere in the peripheral auditory system is information about location available to the owl. The central nervous system apparently computes the location of the sound in space by making comparisons of amplitude and of timing between the responses of the two ears. The map of auditory space in the midbrain nucleus mesencephalis lateralis (MLD) is a true emergent property of the owl's auditory system (Knudsen & Konishi 1978).

Marler (1955) proposed that the thin, high-pitched whistles used as alarm calls by some European song-birds should be hard to locate for their avian predators, while other repetitive, click-like sounds of wide frequency range used as mobbing calls should be easier to localise (see section 2.9.1). He based his arguments on ideas about sound localisation derived from human psychophysical data. He suggested that the alarm whistles should be hard to locate because their frequency was too low for the predator's head to create a significant sound shadow, yet too high for unambiguous localisation using differences between the ears in time of arrival. Marler also postulated that sounds with sudden onset and termination, such as the mobbing calls, should be easier to locate than those having gentle transients, because of the clarity of time-of-arrival cues.

Although Marler's conclusions appear generally correct, birds use considerably different mechanisms to humans (Konishi 1973).

Locating sounds in water

Underwater sound localisation poses problems for mechano-sensory systems. The speed of sound in water is four to five times that in air, thus reducing differences in time of arrival to one-fifth of those in air and increasing wavelengths fivefold. In addition, the acoustic impedance of body tissues often matches that of water, so that to gain any sensitivity at all to sounds, animals must rely upon sensors which are mechanically coupled to an air-filled structure (such as a swim bladder) or to a dense body structure (such as a fish otolith). Experiments on sound localisation are difficult to perform in water because most aquaria are so small that numerous echoes are quickly established, resulting in standing waves. However, Schuijf (1975) and Schuijf and Buduwala (1975) have overcome this by conducting sound localisation training experiments in a deep Norwegian fiord, and have shown that cod can localise sounds of 75 Hz to within 22° in the horizontal direction, and can discriminate between sounds from the front and from the rear. Hawkins and Sand (1977) found even better discrimination than this in the vertical plane, and Myrberg *et al.* (1976) have shown accurate localisation of sound by sharks in field studies.

Electrophysiologists have been able to trace inputs for sound reception in fishes to the inner ear, which consists of the semicircular canals and the otolith organs, the utricle, saccule and lagena. Scanning electron micrographs of the sensory surface of these otolith organs show large zones of remarkably structured, nearly crystalline arrays of hair cells, all polarised in the same direction. The orientation of the hair cells gives the otolith organs sensitivity to particle displacements in particular directions. There is a close correspondence between the orientation of the hair cells and the direction of sounds eliciting the greatest response. Sound inputs from separate otolith organs provide sensitivities to different directions.

Not all sound signals in a given environment are equally locatable. The recipient must make compromises and presumably tunes its auditory physiology to the types of signals most essential to its survival. Given the variety of mechanisms recipients may use to locate signals, it would appear unlikely that signals could evolve which were universally locatable or non-locatable. Rather, as in the case of mobbing calls which are easy to locate and aerial-predator alarm calls which are difficult, signals may have design features which adapt them to a specific group of hearers, in this case the caller's conspecifics and the predatory birds, respectively.

4.6.3   Conclusions about signal localisation

Throughout this section, the emphasis has been on the diversity of signal-localisation mechanisms in animals. Because of the multiplicity of cues available for localisation, it is difficult to generalise about the ease with which a given signal can be located without reference to a specific receiver.

Physical considerations ensure that there are clear differences between modalities in locatability of signals. Light, with its short wavelength, is easily focused or shaded by small structures. The receiver can thus form a two-dimensional map of the visual world, like a photograph, on a surface such as the retina. The precision with which sound can be located in space is far lower than that for vision. Recent results on owls, bats, crickets and amphibians show that accuracy may be achieved, however, by unusual mechanisms. The neural map of auditory space found in the midbrain of the owl is an exciting new finding that may serve as a model for possible localisation systems in other animals. Gradient detection of odours appears to operate only at short range, where gradients are steep. By moving through an odour field or by searching for the motion of the air or water medium, however, odours can be accurately located.

4.7   Conclusions

There is fascination with the study of sense organs because there are often surprises in the way animals view their world. The discovery of unusual organs or the extension of the range of sensitivity into atypical wavelengths or frequencies is always interesting. At first, a sense organ's capabilities may seem obscure and confusing, but as the functions of the organ become clearer, so too does the physiology.

Sense organs are essential to communication. One of the aims of this chapter has been to show how receptor physiology is adapted for communication. In the past, studies of sensory function have been strictly the domain of the sensory physiologist. Although a great deal is therefore known about the peripheral and central structures involved in various senses, our knowledge of the processing of natural stimuli is somewhat limited. This is rapidly changing, however, as more and more ethologists become interested in sensory processing.

Some sensory pathways may have evolved solely for communication sensing. To study the physiological properties of such sensory systems effectively, we have to explore their responses to biologically relevant stimuli, like the ones used in communication. An important new approach to sensory physiology combines the ethological with the physiological in the search to discover how sense organs filter, how they code, how they recognise, and how they locate stimuli.

4.8   Selected reading

Various reference books deal with sensory mechanisms in relation to behaviour, and communication figures strongly in some of these works. The textbook by Marler and Hamilton (1967), now over 15 years old, is still an excellent introduction to the sensory basis of communication. The volume edited by Ali (1978) stresses the adaptive significance of sensory systems, and includes some discussion of communication. Lythgoe (1979) similarly emphasises how vision depends upon environmental conditions, and his book is at its best dealing with aquatic systems. The book edited by Tavolga *et al.* (1981) on the acoustic sense and auditory communication in fishes is an advanced reference work on this specialised system. Reference has been made throughout this chapter to reviews in the *Handbook of Sensory Physiology*. The chapter in it by Land (1981), on the optical properties of invertebrate visual systems, is a highly readable, thorough review of insect visual systems. The volume edited by Busnel and Fish (1980) is an up-to-date reference work on animal sonar systems.

Finally, any discussion of the theory of signal processing eventually enters the field of Fourier analysis. The book by Bracewell (1978) is a relatively advanced treatment of the subject. Bracewell's discussions are mathematically complete, although not totally rigorous. The book stresses subjects of interest to communication engineers, and provides a solid foundation for understanding the mathematics of Fourier analysis.

CHAPTER 5
THE EVOLUTION OF COMMUNICATION:
INFORMATION AND MANIPULATION

R. HAVEN WILEY

5.1   Introduction

Communication occurs when one individual's actions provide a signal that changes the behaviour of another individual. The evolution of communication thus depends on the changes in fitness of the sender and the receiver of a signal. By change in fitness, we mean a change in the rate at which genes influencing an individual's actions spread in the population. This rate depends in turn on the survival and reproduction of individuals carrying these genes. As a result of natural selection in past generations, an individual should only produce signals that increase its fitness. Likewise, an individual should only respond to signals in ways that increase its fitness. This superficially simple situation, however, leads to some fascinating complexities, which are the main subject matter of this chapter.

In recent years, there have been three approaches to understanding how natural selection affects the evolution of communication. The first focuses on the signaller. The main question here is: what strategies for signalling are most effective in evoking responses from a potential receiver? The second approach focuses on ways that signallers and receivers might take advantage of each other. The issue here is: how can a signaller (or receiver) manipulate the behaviour of the other individual to its own advantage? The third approach has relied on the theory of games to analyse how animals should behave in conflicts. In part, this approach asks: how should animals communicate with each other in the course of conflicts? Although aspects of these three approaches have seemed incompatible (Dawkins & Krebs 1978), we shall see that much concordance exists among them. In fact, they illuminate the evolution of communication in complementary ways. This chapter takes up these three approaches in sequence.

156

5.2   Adaptations for efficient communication

5.2.1   Information and noise

The basic phenomenon of communication involves two individuals with signals passing between them. For an objective analysis
of this system, however, we need to introduce a nonparticipant
observer, in our case the ethologist, who monitors the behaviour
and characteristics of the signaller (also called the source or
sender), the receiver and the signal (Fig. 5.1; Shannon & Weaver
1949; Cherry 1966).

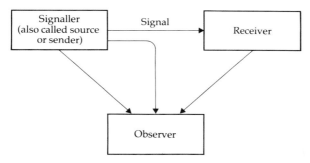

Fig. 5.1. The nonparticipating observer, by recording the behaviour of the signaller
and the receiver and the characteristics of the signal, acts like a privileged receiver
'tapping the wires' of communication between nonhuman animals.

Chapter 2 presented a number of measures of the effects of one
animal's behaviour on that of another, including Shannon's
measure of the information transmitted from a signaller to a
receiver (section 2.11). The term 'transmitted information' has a
technical meaning defined by changes in the predictability of the
receiver's behaviour as viewed by a nonparticipant observer. By
this definition, a signal transmits information to a receiver when
its occurrence increases the predictability of the receiver's subsequent behaviour. An important consequence for the evolution of
communication immediately follows. Whenever the receiver's
response to a particular signal increases the sender's fitness,
selection favours senders that maximise the efficiency of that
signal in transmitting information. In other words, the signal
should evolve to maximise the predictability of the response for a
given time and effort committed by the signaller.

It is important to distinguish between transmitted information and broadcast information. An observer measures the first by an increase in the predictability of the *receiver's* behaviour after a signal occurs. The latter, in contrast, is an increase in the predictability of the *signaller's* identity or behaviour after a signal. Thus broadcast information is a measure of the information obtained from a signal by the observer. It is loosely analogous to Smith's (1968, 1977) concept of the message of a signal. Note that broadcast information accords more closely with the everyday use of expressions like 'this book contains a lot of information'. Yet transmitted information, not broadcast information, is fundamental in an objective analysis of communication, where the primary concern is the effect of a signal on a receiver.

Broadcast information depends on the process of encoding. The signaller translates its internal state into actions or other external changes that produce signals. The internal state of an animal, for our purposes, is the state of its nervous system, which at any moment has a unique 'value' determined by the activity and condition of every neuron. Particular values of an animal's internal state then result in corresponding actions that can serve as signals to others. This mapping of an animal's internal state on to its actions is called *encoding* (see Green & Marler 1979).

The receiver also performs a translation, in this case from the reception of a signal to a change in the subsequent values of its own internal state. These changes might result in an immediate response, but need not. The receiver's translation depends on concurrent external stimulation other than the signal (the context) as well as the current value of the receiver's internal state, which in turn is influenced by its history. This process of translation by the receiver is called *decoding* of the signal.

Intended receivers do not always detect signals nor, even when they detect them, do they always classify them correctly. Furthermore, receivers on occasion respond when they confuse some irrelevant stimulus with a signal of interest. These errors in reception are noise. Note that we have again adopted a technical definition of an everyday word. We usually think of 'noise' as irrelevant sound that masks our perception of something interesting. Noise in the technical sense can indeed result from irrelevant stimulation, acoustic or otherwise, that masks signals, but there are other causes of errors in reception of signals as well.

The degradation and attenuation of signals between source and receiver can also interfere with correct reception (see Chapter 3). After transmission over long distances, two signals can become indistinguishable, or a signal can become indistinguishable from irrelevant stimulation. A bird's song in a forest, at a distance from the singer, is often difficult to distinguish from that of other individuals or even other species and might barely stand out against the noise of wind and rustling leaves.

Furthermore, the receiver's threshold for detecting a signal has a fundamental effect on the errors it makes and consequently on the nature of noise in communication. As mentioned above, there are two kinds of mistakes in detecting a particular signal: missing some occurrences of the signal (missed detections); and reacting to some stimuli that are not the signal (false alarms). A receiver cannot minimise missed defections and false alarms simultaneously (see review by Wiley and Richards (1983)). Instead, it must make a trade-off. A receiver that sets its detectors at a lower threshold for response misses fewer signals but risks making more false alarms. The opposite happens when a receiver sets its threshold higher.

5.2.2   *Improving the detection of signals*

To improve the efficiency of communication, a signaller cannot directly influence the processing of the signal by the receiver's nervous system, but it can employ signals that are easy for the receiver to detect. The theory of signal detection predicts that receivers become more reliable, in other words, miss fewer detection for a given number of false alarms, provided signals have certain features. Four features that increase the reliability of detection are redundancy, conspicuousness, possession of small signal repertoires, and use of alerting components (Wiley & Richards 1983).

Redundancy

Redundancy results from predictable relationships among different parts of a signal. Consequently, a receiver knowing these relationships can identify the kind of signal even if it correctly recognises only a part of that signal. Simple repetition of a signal, or elements of a signal, is the clearest sort of redundancy. However, redundancy can apply to spatial, as well as to sequential, arrangements of the parts of a signal.

The displays of animals have widely varying degrees of redundancy. The more complex and stereotyped the composition

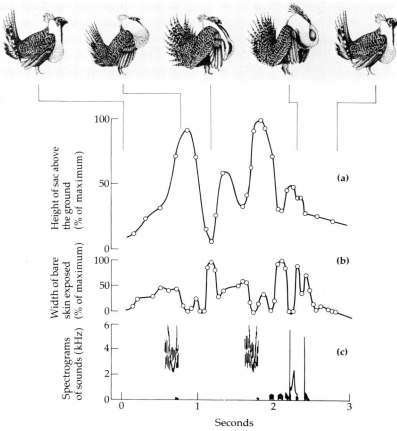

Fig. 5.2. The strut display of the sage grouse lasts nearly three seconds and provides both visual and acoustic signals. **(a)** The strutting male inflates his large oesophageal sac by heaving it upwards and letting it fall twice. **(b)** This manoeuvre exposes bare patches of olive-coloured skin on his chest. **(c)** At the climax of the display, he compresses the inflated sac and then releases the air explosively to produce a ringing pop and low-pitched coos. Since the timing of this display varies only slightly among performances of any one male or among different males, it is one of the most elaborate and stereotyped displays of any bird.

of a signal, the more redundant it is. At one extreme are displays that involve movements of many parts of the body in stereotyped coordinations. The elaborate strut display of male sage grouse (*Centrocercus urophasianus*) provides an extreme example (Fig. 5.2; Wiley 1973). In comparison, the 'push-up' displays of male *Anolis* lizards lack such extreme temporal stereotypy, and some components of the display have low correlations with each other (Jenssen 1971; Stamps & Barlow 1973). In another such case, the song-spread display of Carib grackles (*Quiscalus lugubris*), the elevations of the beak and the wings vary independently (Wiley 1975).

Redundancy in signals has some obvious disadvantages. First, redundancy takes time or requires additional components that could otherwise be used to send more refined messages. Secondly, redundant displays take more time and energy to encode any given message. Even if the response to the signal benefits the sender, why take extra effort or time to produce a redundant signal?

The answer is that redundancy can reduce errors in the detection and recognition of signals. It thus has advantages for a signaller faced with high levels of noise in communicating with a recipient. Noise, leading to errors by the receiver, can reach high levels for several different reasons. First, accurate detection of signals is likely to be difficult during interactions at long range, when signals are often attenuated to near the level of background stimulation or become distorted by degradation in transmission (see Chapter 3). In fact, many long-range signals of animals, such as the advertising songs of male birds or the loud calls of forest primates, are notably complex and stereotyped in comparison with other displays in the same species' repertoire (Marler 1973). In contrast, the signals employed in close-range communication among members of stable social units, like primate troops or mated pairs, have great variability. In the latter case, where redundancy is not so necessary to facilitate accurate detection of signals by recipients, the additional variability can be used to encode further variants of the sender's state.

High levels of noise can also result from the communication signals of other species. Many species of duck that nest in the northern temperate regions form pair bonds on wintering grounds or during migration, when many related species frequent the same locations. In identifying conspecific males, females cannot rely on segregation of related species in different habitats or geographical ranges. Males of these species have conspicuous plumage in elaborate species-specific patterns: this complexity provides redundant visual cues for recognition of species. Males of related nonmigratory species of duck have much drabber plumages.

Other conspecifics can create high levels of noise for mates communicating with each other in dense colonies. Many colonial birds have complex, individually distinctive vocalisations that mates use for individual recognition (White & White 1970; Wiley 1976; Moseley 1979), often in the presence of phenomenal levels of background stimulation from conspecifics using very similar calls.

Another situation likely to result in noisy communication is when a brief but biologically crucial interaction occurs between unacquainted individuals, such as when an animal chooses a mate. In some species, females may have to choose mates rapidly and with little or no prior experience of potential partners, for example when arriving at breeding grounds after migration. When opportunities for heterosexual association are limited, avoidance of errors is vitally important. In such cases, a female should raise her threshold of response to males so as to minimise her chance of making false alarms. As a result, a female would also fail to detect more signals from potential mates. To counteract this problem, males should evolve inherently more detectable signals, ones with greater conspicuousness and more redundancy. It is in birds with brief association of the sexes that plumage and displays of males reach their most extravagant development (Darwin 1871; Sibley 1957). The strut display of sage grouse provides a good example. Such extravagant signals are, from this point of view, adaptations for efficient communication with cautious receivers.

Conspicuousness

The inherent detectability of a signal depends on its conspicuousness, its contrast with spurious stimulation reaching the receiver. Even other types of signal constitute part of the background from which a receiver must distinguish a particular signal. Darwin (1872) recognised this point when he proposed his 'principle of antithesis' in behaviour: actions accompanying contrasting 'emotions' (values of internal state, in our present terminology) often have contrasting form (section 1.2.2). The examples in the preceding section reveal that conspicuousness of signals often accompanies complexity and stereotypy, aspects of redundancy.

Small repertoires and typical intensity

The theory of signal detection predicts that receivers do better, in the sense that they miss fewer detections for a given rate of false alarms, when they must classify signals into fewer categories. This theoretical prediction has been confirmed in psychophysical experiments on human subjects (reviewed in Wiley & Richards 1983). The smaller the repertoire of signals that a receiver must identify, the better is its performance. The results apply with great generality to signal detection, regardless of specific mechanisms, to animals as well as to electronic devices.

Small repertoires for communication in potentially noisy circumstances could explain variation in repertoire size among North American species of wren. In those species with dense populations in habitats occupied by few other passerines, such as marshes or desert, individuals have large repertoires of song patterns, sometimes over 100 (Kroodsma 1977). In comparison, individuals of those species with sparser populations in habitats with diverse avifaunas, such as forest or broken woodland, have relatively small repertoires. In this case, the greater average distance between territorial neighbours and the presence of additional species tend to make communication noisier.

The advantages of a small repertoire can also explain a striking feature of many displays—'typical intensity'. Displays with a typical intensity maintain a standard form over a range of a signaller's internal state, instead of varying with it (Morris 1957); typical intensity thus reduces information broadcast in a display. Fewer and more-distinct signals as a result of typical intensity in displays reduces ambiguity (Cullen 1966), but a clearer way to understand the effects on a receiver is to focus on the reliability of signal detection.

Rather than producing signals which vary in a complex way, in an effort to evoke varying responses from a receiver, it would pay when noise is a problem to produce one or a few standardised signals that would have a greater chance of reliable detection. Consider a situation in which a signaller would benefit from a prompt response by the receiver, perhaps recognition by its mate; promptness of correct recognition might be more important than the exact form of the response. The signaller would have to strike a balance between uttering a standard signal that would have greater chance of quick, correct recognition and uttering one of a number of possible signals that could evoke a more specific response but which might need repeating before correct recognition occurred. Typical intensity thus serves to improve the accuracy of detection and recognition of signals by a receiver.

Alerting signals

An alerting component at the start of a signal assists a receiver by specifying the interval of time during which it can expect to receive the remainder of the signal. The alerting component must have high inherent detectability, in other words little degradation during transmission and maximum contrast with the background. On the other hand, it need not encode much information about the signaller. For instance, it need not permit recognition of the individual or even of the species producing the signal. The subsequent message component, on the other hand, might well encode information about the identity and internal state of the sender. The receiver can thus set a relatively high threshold for response and still have a satisfactory level of correct detections in relation to false alarms. Once the alerting signal is detected, the receiver can then lower its threshold for the precise interval of time that the message component occupies. By knowing the time of onset of the message component, the receiver can detect and recognise it more reliably. Alerting signals thus permit a receiver to devote less time to being attentive to signals and more to other activities, such as foraging, without any significant reduction in its vigilance. From the sender's point of view, an alerting component increases the chances that a receiver will detect and recognise the message component of a signal (Raisbeck 1963; Wiley & Richards 1983).

Vocalisations used in territorial behaviour are signals that are produced intermittently but to which animals need to be constantly attentive. The territorial songs of many birds and the loud calls of some forest primates begin with a single tonal component and then become more complex in their acoustic structure. The introductory note seems ideally suited as an alerting component. Field experiments with rufous-sided towhees (*Pipilo erythrophthalmus*) have shown that the initial tonal component of the song (Fig. 5.3) evokes little response but does permit more reliable responses to subsequent, more complex components (Richards 1981b).

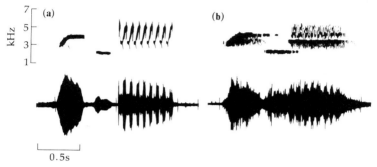

Fig. 5.3. Spectrograms (displays of sound frequency versus time) and oscillograms (displays of sound amplitude versus time) of a typical song by a male rufous-sided towhee illustrate the degradation of acoustic structure by reverberations which occurs during transmission of the song through woodland. The distinct notes of the song at the source (**a**, left) become run together by reverberation (**b**, right). The two clear tones that introduce the song are less severely degraded, however, than the trill of rapid glissandos at the end. These introductory tones serve as an alerting compoment, which serves to call a listening towhee's attention to the message component that follows. (From Richards 1981b.)

5.2.3   Conclusion: Ritualisation as an adaptation for efficient communication

A single display often incorporates several adaptations for improving a receiver's detection and recognition of signals. Redundancy by repetition and stereotypy, contrast with the background, and typical intensity often go together. Ethologists describe a display that shows these features as 'ritualised' (see section 2.2). These adaptations benefit a signaller by counteracting noise in communication. Noise, in the technical sense, can result from cautious receivers as well as from high levels of irrelevant stimulation or from degradation of signals in the external environment.

5.3   Communication as manipulation

5.3.1   *Selfishness of signallers and receivers*

So far we have concentrated on the ways in which a signaller can improve the effectiveness of a signal in evoking a response. The consequences of the response for the receiver's fitness must also be considered in an analysis of the evolution of communication. In general, it is clear that selection should favour responses to signals that raise the receiver's fitness, just as it favours production of signals that raise the signaller's fitness. Individuals, both when signalling and when responding, should tend to act in ways that increase their fitness.

Dawkins and Krebs (1978) developed a case for 'manipulation' of receivers in ways that increase the fitness of the signaller but not that of the receiver. Signals should not evolve, they argue, to 'provide information' to receivers but to induce them by any means possible to behave in a way that benefits the signaller.

Clearly, receivers should also evolve such 'selfish' tactics. They should not necessarily respond as 'directed', but should use information derived from the signaller's acts, in any way possible, to increase their own fitnesses. We can classify the four possibilities according to whether an association of signal and response increases or decreases the signaller's and receiver's fitnesses, (Table 5.1).

Signallers can manipulate receivers by employing deceit: signals conveying something incorrect about the signaller. If a male competing for mates could indicate that he was larger than

Table 5.1. A 2×2 classification of communication based on Hamilton's (1964) classification of social interactions in general.

		Change in receiver's fitness	
		Increase	Decrease
Change in signaller's fitness	Increase	Mutuality	Deceit (manipulation by signaller)
	Decrease	Eavesdropping (manipulation by receiver)	Spite

his actual size, he might gain an advantage over his rivals; of course, his rivals would have to be fooled. The deceitful signaller in effect exploits the receiver's rules for decoding signals. If the existing rule relates a particular size of horns, for instance, to males of a certain level of fighting ability, then a male might gain by growing larger-than-normal horns and bluffing a rival. Alternatively, if the existing rule relates a certain plumage to females or young that do not compete for territories, a male might benefit by adopting this plumage and entering rivals' territories unchallenged for surreptitious feeding or even copulating. The possibilities for deceiving receivers thus depend on the receivers' current rules for translating signals into their responses. These rules need not, of course, be consciously recognised; they could result from any decoding mechanism that associates external stimulation from communication signals with particular responses.

Receivers can manipulate signallers by obtaining information about the signaller against its own best interests. Eavesdropping is a clear case: signals intended for one receiver are intercepted by another. For instance, a male's displays in courting a prospective mate might well attract rivals as well. In addition, it is reasonable to suppose that receivers might take advantage of any imperfection in deceit. After all, deception succeeds only when the signaller does not, in some other way, reveal its true nature. If such perfect control fails, then receivers have some chance of 'reading' the signaller's true state in spite of its attempts to mislead the receiver.

Thus manipulation in communication cuts both ways. Signallers might, in some circumstances, manipulate receivers as a result of constraints on the latter's responses to signals. On the other hand, receivers might also manipulate signallers as a result of constraints on the latter's production of signals.

In thinking about opportunities for manipulation in animal communication, analogies drawn from human interactions tend to dominate. For this reason, it is important to pay special attention to terminology. As in our previous discussions of information and noise, we have provided technical definitions for everyday terms, like 'deceit' and 'selfishness'. These familiar words make visualisation of technical discussions easier, but we must always guard against misleading inferences that can result from loose usage of technical terms.

Manipulation in communication between members of the same species also has parallels with some well-known interactions between species. Thus, deceit in intraspecific communication is paralleled by various forms of mimicry, bluff and feigning by prey in response to predators. On the other hand, predators and parasites often eavesdrop on signals inadvertently produced by their prey, or 'read' the vulnerability of individual prey from their actions or appearance. These analogies with human interactions and with predator–prey or parasite–host interactions suggest two considerations important for any study of manipulation.

5.3.2   *Relative rarity of deceiving signals*

First, misleading signals must occur only rarely in relation to correct ones. In other words, signals following the prevailing rule for encoding and decoding must predominate. Thus in Batesian mimicry, in which a palatable species of prey gains some protection from predation by resembling a distasteful or poisonous species, the mimics must occur infrequently relative to the models. Otherwise, predators would not reliably learn to avoid the mimicked stimulus. If receivers do not encounter misleading signals sufficiently infrequently in relation to correct ones, they should readjust their rules for decoding signals.

This conclusion needs some refinement, however, since in addition to the probabilities of each kind of event consideration must be given to the consequences for the receiver's fitness of responding to misleading signals on the one hand, or failing to respond to correct ones on the other. Imagine a fox that occasionally encounters a plover fluttering one wing on the ground. The plover might have a broken wing and thus provide a meal if the fox could catch it, or it might have a nest nearby and only feign a broken wing in order to distract the fox from the nest.

Should the fox adopt the rule 'stalk the bird', or the alternative, 'look for a nest', when it encounters an apparently crippled plover? Suppose the value of stalking, provided the plover really is crippled, equals $p_s \times w_s$, the probability of success when stalking times the amount of food obtained if the stalk is successful. Similarly, let the value of searching for a nest, provided the plover is pretending, equal $p_n \times w_n$, with symbols analogous to the first case. Finally, suppose a proportion f of fluttering plovers are really crippled, and the rest, $1-f$, are deceitful. Now, the yield to a fox that adopts the rule to stalk is $f \times p_s \times w_s$; the yield to one that adopts the rule to search for a nest is $(1-f) \times p_n \times w_n$.

Which alternative has the greater yield and thus will tend to confer the greater fitness? Stalking will be superior to nest-searching provided that

$$fp_sw_s > (1-f)p_nw_n$$

If $p_s = p_n$ and $w_s = w_n$, then f must exceed one-half for stalking to pay. In other words, more than half the fluttering plovers encountered must really be crippled. Conversely, sham fluttering only deceives foxes, in the long run, provided it occurs in the presence of foxes less often than the real fluttering of cripples. If $p_s \neq p_n$ or $w_s \neq w_n$, as is likely, other solutions result.

5.3.3 Manipulation and information

A second general point that needs emphasis is the distinction between information and manipulation. We have so far provided exact definitions of 'information' and 'manipulation'. In the tripartite system of an ethologist studying animal behaviour, with a signaller, a receiver and a nonparticipant observer, the transmitted information depends on the association between a signal and the receiver's behaviour. There is no distinction here between 'correct' and 'deceitful' signals. The observer can, in addition, determine the broadcast information in a signal about the sender's state or identity and could thus compare the broadcast information and the transmitted information, to determine whether or not the receiver uses the information available in a signal. The observer could also determine whether signals conceal important features of the signaller, by bluffing or mimicry for instance.

Manipulation, in any of its forms, depends entirely on the relative changes in the fitnesses of the signaller and receiver. It is independent of the definitions of transmitted and broadcast information. We have emphasised that selection should act on senders to increase the efficiency of transmitting information whenever the sender's fitness is increased by the response. Selection should act just as well on receivers to minimise this efficiency whenever the receiver's fitness is reduced.

In practice, many difficulties arise, not only in estimating amounts of information but also in estimating changes in fitness. Yet it is important to recognise that transmitted information, broadcast information and manipulation are three distinct features of communication. All three can in principle be estimated for any set of communicatory interactions.

5.4   Evolution of deceit

5.4.1   *Deceit by signallers, retaliation by receivers*

Mimicry or bluffing by signallers, as we have seen, depends on the receiver's rules for decoding signals into responses. Deceit occurs in effect when signallers can take advantage of the receivers' rules. Thus receivers can retaliate by a change in their rules. Two related possibilities exist:
(1) a devaluation or recalibration of the association of signals with responses; and
(2) use of supplementary signals for finer discrimination of the states or identities of signallers.

Inflation and devaluation of signals

Bluffing, deceit by inflation of a single cue (or a correlated complex of cues) for a response, leads to selection pressure on receivers to devalue the cue. If the size of antlers, for example, is a cue for overall size and fighting ability in rival males, then any bluffing by signallers which evolve larger-than-normal antlers stimulates selection for receivers to readjust their rule for decoding the size of antlers. Such a process tends to accelerate once started (Dawkins 1976b; Dawkins & Krebs 1979). After devaluation of a cue, all signallers that do not bluff are placed at a disadvantage. Furthermore, as the devaluation spreads among receivers, selection favours further bluffing, greater inflation of the cue. The escalation of signals and decoding finally ceases when further inflation of the cue becomes too costly or risky for signallers. In other words, the increased risk of predation or loss of opportunities to feed, for example, just balance any beneficial effects of bluffing on the signaller's fitness.

This sort of escalation of bluffing and devaluation could explain why many territorial birds have large repertoires of song patterns

(Krebs 1977). Suppose intruding males seeking openings for territories do best to avoid areas with high densities of established
birds. If they judge the density of established birds in an area by
the number of song patterns heard there, then resident males
could bluff by each singing more than one song pattern. Escalation
of the sizes of males' repertoires would ensue, as intruding males
devalue the diversity of song patterns as an index of the density of
established territories. More inflated bluffing would lead to more
devaluation until the acquisition of larger repertoires became too
costly, perhaps in terms of the time required for learning, in
comparison with the benefits from fooling intruders.

Some experiments provide initial support for this idea. When
male great tits (*Parus major*) are removed from their territories,
reoccupation of the resulting vacancies by new males is delayed
by tape-recordings broadcast in these areas (see section 1.4.1).
Recordings that include many song patterns have a greater effect
than those with fewer, suggesting that the number of song
patterns heard in an area does indeed influence males deciding
where to set up territories (Krebs *et al.* 1978; Yasukawa 1980). It is
not clear yet how the sizes of repertoires influence intruders'
behaviour. Repertoires might indicate the probable density of
established territories or the probable fighting capabilities of individual territorial residents, or particular song patterns might differ
in effectiveness (Krebs & Kroodsma 1980). The fact remains that
any of these possibilities invites bluffing and consequent escalation of signals.

In the end, such escalated signals would no longer be deceptive
once devaluation by receivers completely compensated for inflation of the signal by senders. Such costly, but no longer deceptive,
signals might conceivably fall into disuse, since selection might
favour substitution of less costly signals. These new signals would
then lead to a new round of escalation by inflation and devaluation
(Andersson 1982). On the other hand, there are some reasons to
expect that selection might maintain costly honest signals, as we
shall see below.

Increased discrimination by receivers

Bluffing and mimicry need not lead to escalation in the cost of
signalling, however, when receivers can adopt countermeasures

other than devaluation. Another way for receivers to counter mis-representation is by increased discrimination among signals. Rather than devalue an unreliable cue, receivers should instead attend to additional cues or scrutinise signals in more detail. For example, some probing of a signaller's reactions might reveal its true mettle, even when some try to bluff. Of course, each additional cue or detail that a receiver examines is susceptible to the same possibility of escalation by inflation and devaluation. In some cases, this process would lead to exact imitation of a model by its mimic. However, particularly in intra-specific communication, the effects of multiple assessments by receivers presumably stop short of exact mimicry by deceitful signallers. After all, carried to the extreme, exact mimics must become indefinitely similar to the model. We must seek limits to the advantages of mimicry by signallers or, conversely, limits to the advantages of discrimination by receivers.

In some cases, there may be no advantage to signallers in indefinitely exact mimicry. Consider sparrows that compete for food in winter. In some of these species, younger individuals have duller plumage than older birds. An extreme example is Harris' sparrow (*Zonotrichia querula*) in which older males in winter have much more extensive black patches on their throats and breasts than do females and younger males (Fig. 5.4). Since older males tend to dominate other birds in contests over food, the size of a bird's black patch is a cue for its status. Experiments show that dying a pale bird's breast black, in order to increase its resemblance to a dominant male, also increases its success in competition with other subordinate birds (Rohwer 1977; section 1.4.3). Why then do not females and young males bluff by evolving larger black patches? The interactions of the disguised subordinates with dominant males provide an answer. The older, dominant birds persecute the dyed individuals. Evidently, they detect the disguised birds by their behaviour, perhaps by their reactions when challenged. If the disguised birds are also treated with the male hormone testosterone, then they can rise in dominance to equal even the true dominants (Rohwer & Rohwer 1978). Bluffing in this case would require changes in a signal indicating dominance, an extensive black patch, but also adoption of the behavioural reactions typical of a dominant. If the costs of acting like a dominant are too great in relation to the benefits so derived by a young individ-

Fig. 5.4. The throat and breast feathers of Harris' sparrows in winter are unusually variable in colouration. The birds with more extensive black patches tend to dominate those with less black. Although the variation is almost continuous, older males tend to have more black than younger males and females. By painting some pale birds' throats black, Rohwer (1977) could create individuals that mimicked more dominant birds.

ual, it would do better to avoid bluffing that requires such exact mimicry. In general, if the costs and benefits of achieving a particular social position differ for two individuals, it might not pay the inferior to mimic the other so closely that it must adopt most of its characteristics.

Increased discrimination by receivers might also incur disadvantages. A limit on discrimination would then permit the evolution of deceit by partial mimicry. Cases of males mimicking females, in order to take advantage of rival males, are good candidates here (see section 2.8.3). For instance, young male elephant seals (*Mirounga angustirostris*), like females, are smaller than the older males that defend harems. By behaving like females and mixing with the harems (Fig. 5.5), these young males can try to sneak copulations (Le Boeuf 1974). Although the harem masters expel them when detected, the level of vigilance and aggression necessary to exclude all sneaking males could well disrupt the harem enough for the master bull to lose many of the females. In these circumstances, a harem master should increase his level of discrimination (both detection and eviction) until any further increase results in more copulations lost as a result of disrupting his harem than gained as a result of evicting males that mimic females. An interesting complication here is provided by the females. By producing loud screams during copulation, especially with smaller males, they ensure that only the most dominant male within hearing normally completes copulations (Cox & Le Boeuf

Fig. 5.5. A harem of elephant seals illustrates the predicament faced by the harem master, an old bull (centre, with his head and large proboscis raised). He attempts to guard the numerous females (smaller and lighter in colour) from the attentions of other males, including several younger males that hang around the periphery of the harem (somewhat larger and darker than the females, left edge and right background). Nevertheless, one young male has infiltrated the harem (right foreground). In response to a threatening call from the female next to the interloper, the harem master prepares to roar a warning. (Photograph by B.J. Le Boeuf.)

1977). In effect, they make the harem master's task of detecting and thwarting sneaky males easier.

Another example of males mimicking females is provided by the scorpionfly (*Hylobittacus apicalis*), in which males present females with a nuptial meal of a dead insect before copulation (Thornhill 1979). Sometimes a male hunts for and captures prey itself, but at other times he steals prey from other males. In the latter case, the thief sometimes mimics a sexually receptive female in such a way that the rival male gives up its prey. This ruse does not always work, as the duped male occasionally discovers his mistake and snatches the prey back. Here a male carrying prey must balance the advantages of increased vigilance against thieves against the advantages of responding quickly to receptive females. Too much probing of the credentials of apparent females might lose him prospective mates, a consequence of courting too slowly, while insufficient probing would lose him his nuptial offerings, a consequence of gullibility to female mimics.

In these cases of males mimicking females, receivers (the seal harem masters or the scorpionfly males carrying prey) are caught in a double bind: discrimination of deceivers from females has conflicting advantages and disadvantages. The net benefit of any level of discrimination or probing depends on the relative frequencies of mimics and models encountered. Probing should increase as the frequency of mimics rises and decrease as it falls.

5.4.2   Unbluffable signals

Bluffing, as we have seen, confers a disadvantage both on the gullible receiver and also on the honest signaller. We have considered how the receiver might evolve counteracting adaptations. Those honest signallers that serve as models might also evolve measures to avoid their exploitation by bluffers. For example, suppose size has an effect on the outcomes of fights. Then the largest individuals, which are exploited by the bluffing of smaller ones, would gain by evolving 'unbluffable' signals indicating size. For instance, a deep voice might be an unbluffable cue for size in toads (*Bufo bufo*) and other animals (Davies & Halliday 1978; Morton 1977). This is because the pitch of an individual's voice generally correlates with the size of its sound-generating structures in the larynx or syrinx, which in turn correlates with overall size. Phylogenetic or physiological constraints, one must assume, prevent evolution of the larynx or syrinx independently of the rest of the body.

Perhaps the clearest possibility for the evolution of unbluffable signals comes from the escalation of signals, by counteracting inflation and devaluation, to produce a signal that has become too expensive to inflate further. Signals should evolve by this process to become as expensive as the benefits permit. Honesty would accordingly entail a high price. It has even been suggested that signals should evolve to become a net handicap to signallers (Zahavi 1975), although this proposal seems unsound (Davis & O'Donald 1976; Maynard Smith 1976a).

5.4.3   *Conclusion: When can deceit persist?*

This section has compared three consequences of deceit in animal communication: escalation by counteracting inflation and de-valuation of signals; mimicry limited by disadvantages of inci-dental consequences for the signaller; and mimicry limited by disadvantages of increased discrimination by the receiver. Selec-tion on signallers in some cases favours inflation of signals by bluffing, but in other cases favours the use of unbluffable signals. Selection on receivers faced with inflated signals favours devalua-tion or increased discrimination of signals, although there are sometimes limits to the advantages of increased discrimination.

The outcomes of the three scenarios listed above are not easy to predict exactly. The first, escalation, would often lead to costly but honest signals. Once bluffing became too costly, no further deceit would occur. The second scenario for the evolution of deceit, mimicry limited by disadvantages to the signaller, would also lead to universally honest signals, provided receivers could match any deception with increased discrimination. In contrast, the third scenario, mimicry limited by disadvantages of increased dis-crimination by the receiver, results in the indefinite persistence of deceitful signals. The advantages of increased discrimination tend to increase as mimics increase in frequency and to decrease as mimics decrease. The level of discrimination by receivers reaches a stable compromise depending on the relative frequency of deceivers encountered by a receiver.

5.5 Communication in contests

5.5.1 Withholding information

Withholding information about identity or internal states is sometimes the best course to take. Clearly, signallers attempting to deceive receivers should withhold as much information as possible about their true condition. Likewise, signallers susceptible to eavesdropping by competitors or rivals should direct signals as narrowly as possible to intended receivers.

Theoretical analyses of strategies for fighting also suggest that individuals should often withhold information about themselves. It is important to emphasise that 'information' as used here is broadcast information, which an observing ethologist might receive. In fights, individuals are of course selected to act in ways that influence their opponents' behaviour, in other words to transmit information, as effectively as possible. Broadcast information is another matter, though. Recent analyses of the actions of birds and fish in competitive interactions suggest that displays in contests are in fact generally poor predictors of an individual's following actions or the outcome of a fight (Caryl 1979). In other words, the level of broadcast information for these displays is low.

One possible interpretation of such findings is that ethologists have not yet detected all of the broadcast information, since it is likely to be complex or subtle (Hinde 1981; van Rhijn 1980). It seems clear that competing individuals should never volunteer unconditional information about their next move, except perhaps before surrender or retreat. The element of surprise or the possibility of negotiation preclude any advantage for signals providing unconditional information. The analogy with human combat and diplomacy is persuasive.

Providing more complex information, on the other hand, could well have advantages. Particularly likely here is information about contingent behaviour: a signal that indicates the likelihood of particular responses following a move by the receiver. Possibilities include signals that indicate commitment to retaliate if the receiver attacks or to withdraw if the receiver withdraws. A signal indicating 'I will fight if and only if attacked' would particularly assist an individual defending a territory or mates. Such complex predictions about a signaller's future actions have possibly escaped the attention of ethologists.

5.5.2   *Bluffing as an initial strategy*

Another possible explanation for the apparent lack of broadcast information in contests is bluffing. In human bargaining, in the absence of perfect information about the opponent's intentions and resources, it usually pays to demand more than one expects from an opponent or to threaten more than one can deliver. The same seems reasonable in contests between nonhuman animals (Maynard Smith 1974; Maynard Smith & Parker 1976). Individuals should initiate confrontations with threats of maximum intensity, within the bounds of the receivers' gullibility.

This case nicely fits our third explanation for the persistence of deceit: disadvantages of increased discrimination by the receiver. At least when contestants differ only slightly, time presumably limits possibilities for accurate discrimination of abilities at the outset of contests. Rapid judgments of an opponent carry risks of unnecessary withdrawal or premature attack, the latter with consequent chances of injury. Thus mutual bluffing and cautious response make good initial strategies and would result in little broadcast or transmitted information. Once contestants begin to feel each other out, in other words learn more about each other's capabilities, then a contest can proceed to a resolution. Contests between female Siamese fighting fish (*Betta splendens*) fit this pattern (Simpson 1968).

The problem of assessing opponents' capabilities should be greatest when contestants are most similar; consequently, contests between well-matched individuals should require the longest time for settlement. Observations of fighting animals tend to support this expectation (Riechert 1978; Sigurjónsdóttir & Parker 1981).

Bluffing at the outset of a contest also provides an explanation for typical intensity of displays (Maynard Smith 1974; Maynard Smith & Parker 1976). In fact, this explanation and our earlier one, based on adaptation for efficient signal detection in a noisy environment, do not conflict. As noted, an opponent at the outset of a contest is likely to respond cautiously, introducing noise to the communication channel (section 5.2.1). Signals in this case should evolve redundancy, conspicuousness and typical intensity.

5.5.3 Uncorrelated asymmetries

Is there ever an advantage to contestants that respond to signals conveying no information whatever about the sender? In theory, this is a possibility. Imagine encounters between two individuals that are indistinguishable so far as they can themselves detect, except for a clear difference that has no relation to their respective fighting abilities or their rewards for winning or losing. Such a difference is termed an 'uncorrelated asymmetry'. Suppose further that any individual is equally likely to have either of the arbitrarily determined characteristics, A and B. In such encounters any two opponents do best to decide the outcome solely on the basis of this difference, by an arbitrary rule that the animal characterised by either A or B wins (Maynard Smith 1974, 1976b). Thus a signal with no information about the state of the sender would still have a predictable effect on the receiver, in the long run to the mutual advantage of each contestant. The obvious human example of this kind of behaviour is tossing a coin to decide a contentious issue.

It has been suggested that residence on a territory could provide such an arbitrary signal, with 'owner wins' the arbitrary rule for deciding encounters between owners and intruders. It seems doubtful, however, that ownership of a territory is often a completely arbitrary asymmetry. The stronger or older individuals often get territories in the first place, or residence on a territory might increase an individual's size or strength by providing a reliable source of food, or a territory might become more useful with time as a resident learns the terrain. In each case, ownership would correlate with superior fighting abilities or greater rewards for winning. It is likely to be difficult to determine whether or not fights between animals are ever decided solely on the basis of an arbitrary signal, but one species in which they seem to be is the speckled wood butterfly (*Paraage aegeria*), contests over occupation of territories being settled by an 'owner wins' ruling (Fig. 5.6). The territories defended by these butterflies are of very little value; they contain no resources and, being patches of sunlight, they are highly ephemeral. Furthermore, the delicacy of their wings would make any more severe form of fighting very costly (Davies 1978).

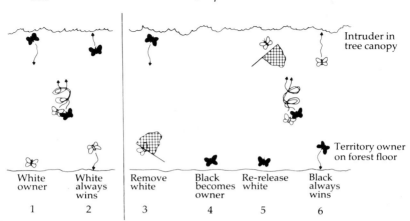

Fig. 5.6. Territorial behaviour in the speckled wood butterfly (*Paraage aegeria*).
1 and 2: when an intruder (black) enters a territory, there is a brief interaction
involving a spiral, upward flight, at the end of which the resident (white) always
retains his territory. 3–6: if the original resident is experimentally removed, a new
butterfly quickly becomes the resident and subsequently wins contests with
intruders. (From Davies 1978.)

5.5.4   Contests with no information

When opponents lack any information about each other, even
about arbitrary differences, some clear hypotheses result. Imagine
contests that are settled by which opponent persists longest. The
opponents' costs, we can assume, increase in proportion to the
duration of the contest. Suppose that opponents, who know
nothing of each other's intentions, select their duration of
maximum persistence in advance of each contest. Then whoever
has selected the longer time wins. This is the 'war of attrition'
(Maynard Smith & Parker 1976; Maynard Smith 1974, 1976).
Clearly it does not pay for an animal to adopt a strategy of fighting
for the same duration in all fights. It would always lose to individuals that fought longer. It turns out that an evolutionarily stable
strategy, one that cannot be invaded by mutants with any other
strategy, involves selecting a maximum persistence (and a corresponding maximum cost) from a negative exponential distribution.
Either each individual can select a new persistence for each fight
or individuals with the same genotype can nevertheless differ

consistently in persistence, so long as persistences fit the required distribution.

This argument leads to the prediction that the duration of fights, like the preselected persistences, should fit a negative exponential frequency distribution. In fact, the durations of fights in several species do fit this prediction. One might be tempted to conclude that animals fight wars of attrition on the basis of no information about opponents. However, in a carefully analysed case the evidence does not support this conclusion. In fights between male dung flies (*Scatophaga stercoraria*) over females (Parker & Thompson 1980; Sigurjónsdóttir & Parker 1981), the durations of fights fit the predicted distribution, but this is for some other reason. There is clear evidence that opponents do acquire information about each other. Smaller males usually withdraw, for instance, and prior possession of a female also confers an advantage.

If a complete absence of broadcast information in signals seems improbable, even in fights, signals that completely specify opponents' characteristics are equally unlikely. The problem is that animals fight too much for this possibility to hold up. If opponents have complete and accurate information about each other's fighting abilities and the rewards from winning or losing, then no fight should ever occur; the inferior or less-committed animal should withdraw before any contest escalates to a fight (Parker 1974; Maynard Smith 1979; Parker & Rubenstein 1981). Animals clearly do fight, often frequently and severely.

In view of this discussion, it seems likely that animals often do obtain information about opponents during contests, although initially only incomplete or contingent information. Opportunities for bluff are also clear, but otherwise the role of signals that broadcast no information about the sender remains elusive.

5.6   Evolution of simple honesty

5.6.1   *Mutuality and competition*

Communication, we have seen by now, takes both mutualistic and competitive forms. In some situations, it has advantages for both sender and receiver; in others, one can take advantage of the other. Honesty in signalling can result from competition, as discussed above, either by counteracting escalation and devaluation of signals or by limitations on the advantages of deception for signallers. These forms of honesty require either high costs for escalated signals or special limitations on deception. What about simple honesty? Can signals evolve so that they broadcast information about the signaller without escalation or special limitations on deception? For the evolution of this form of honesty, mutualistic cooperation rather than competition provides the most straight-forward explanation. It is thus important to consider the circumstances which favour the evolution of mutuality in communication. In particular, we must ask whether mutuality can evolve even when manipulation has advantages for the signaller or receiver, as so often happens in communication.

Kin selection provides one way in which mutuality in communication can evolve. Kin selection results from the presence of the same genes in related individuals, by virtue of their descent from a common ancestor. An individual that manipulates a close relative to increase its own survival or reproduction also has a certain probability of reducing the survival and reproduction of the same genes in the relative. Manipulation of relatives is thus not always favoured by selection, depending on how close the genealogical relationship is and on the benefits to the signaller and the costs to the recipient.

5.6.2   The Prisoners' Dilemma

In the case of unrelated individuals, evolution of mutuality is best analysed by means of game theory. In such analyses, the net benefits to an individual (expected changes in its fitness) depend both on its own strategy and on the strategy played by its partner. Our discussion has emphasised how characteristic this dependence is for communication. The benefits of deceit depend on whether receivers are gullible or discriminating. The benefits of devaluation or discrimination depend on whether or not the sender is deceitful or honest.

When manipulation has advantages for signallers or receivers, it seems at first that mutual cooperation, simple honesty in signalling and trust in receiving, cannot persist. A population comprising only honest signallers and trusting receivers invites invasion by mutant individuals with tendencies for deception,

Table 5.2. The pay-offs for two individuals, A and B, who can adopt either of two strategies, cooperating and defecting. In each cell, the first symbol represents the pay-off to A, the second that accruing to B.

		Individual B Cooperating	Defecting
Individual A	Cooperating	P, P	S, R
	Defecting	R, S	T, T

since deceitful signallers gain in interactions with trusting receivers. This deceit in signals creates an advantage for discriminating or devaluing receivers. Unless there are limitations to the advantages of deceit or discrimination, as discussed above, a population would eventually contain only mimics and discriminators or bluffers and devaluers. Such a population could not be invaded by mutant individuals with tendencies for either honesty or trust.

Yet there is a dilemma. The net benefits to individuals in a population of deceivers and sceptics are lower than those in a population of honest and trustful individuals. This situation suggests comparison with the Prisoners' Dilemma of game theory, named after the situation facing hypothetical prisoners who can escape only by cooperating with each other but who can obtain leniency by defecting to report their comrades to the authorities.

The Prisoners' Dilemma permits formal treatment as follows. Imagine two strategies, Cooperating and Defecting. Each individual can select either strategy, either permanently or for each separate interaction. The rewards to an individual then depend on both its own and the partner's strategy. Each cell in Table 5.2 presents the reward (called pay-off in game theory) for two individuals. Note that the players have symmetrical roles: when one defects and the other cooperates, the pay-offs to the defector and cooperator are the same regardless of which individual plays which role.

The Prisoners' Dilemma exists when $R > P > T > S$. When others cooperate, it pays to defect ($R > P$). When others defect, it still pays to defect ($T > S$). Yet the rewards of defecting among other defectors are not so great as those of cooperating with other cooperators ($P > T$).

An evolutionarily stable strategy for this game is to defect. No matter which strategy the partner selects, an individual that defects always does at least as well as its partner. In fact, whether this game is only played once between any two partners, or whether it is played any predetermined number of times, defecting appears to be the only evolutionarily stable strategy (Rapaport 1960; Rapaport & Chammah 1965; Axelrod & Hamilton 1981).

The situation changes, however, when partners play an indeterminate number of times. When there is always a possibility for at least one more interaction between any two partners, the strategy called 'tit-for-tat' can prevail. This strategy is a conditional one: in any particular interaction, an individual cooperates if its partner cooperated in the previous interaction, but defects if its partner defected previously (Axelrod & Hamilton 1981). Thus, particularly if partners occasionally test each other's willingness to cooperate, any two partners tend to lock into cooperative interactions and avoid the pitfall of mutual defection.

This result has two prerequisites: individuals must remember their partners in interactions, and any two partners must interact an indeterminate number of times. Such conditions are probably met by many communicating individuals. Among birds, territorial neighbours and long-term mates are obvious candidates. Individual recognition is well documented for both these cases in birds (Falls & Brooks 1975; Wiley & Wiley 1977; Jouventin *et al.* 1979; Moseley 1979). Communication within stable social groups, such as primate or canid groups or birds that defend group territories, could also clearly meet these criteria.

5.6.3   *Cooperation and defection in communication*

How closely, though, does deceit in communication fit the basic model for the Prisoners' Dilemma? In communication, signaller and receiver do not always have symmetrical roles. It is true that roles can change reciprocally, as often occurs in interactions between territorial neighbours. Other cases of communication, for instance between mates, do not always have even such long-term reciprocity. Is strict symmetry necessary for the preceding analysis to hold?

Imagine a game of communication with interactions between a signaller who can produce either honest or bluffing signals and a receiver who can either trust or devalue these signals. By analogy with the example above, we can tabulate the pay-offs from each of the four possible kinds of interactions as in Table 5.3.

Table 5.3. The pay-offs for two communicating animals, one of whom can be an honest or bluffing signaller, the other a trusting or devaluing receiver. In each cell the first symbol represents the pay-off to the signaller, the second that accruing to the receiver.

| | | Receiver | |
		Trusting	Devaluing
Signaller	Honest	P, P'	S, R'
	Bluffing	R, S'	T, T'

Here we have distinguished between pay-offs that we previously assumed to be equal. In the present game, signaller and receiver do not necessarily have symmetrical roles. Hence the receiver's pay-offs are identified by primes.

How are these pay-offs related for each individual? Take the signaller first. We can easily imagine that $R > P$: deceit pays when your opponent is trusting. Furthermore, $P > S$: an animal is at a disadvantage when opponents devalue its signals, for example by reacting as if the signaller is smaller than it actually is. Also, $P > T$, provided bluffing signals are more costly or risky than honest ones, as discussed above. Finally, $T > S$, since signallers would presumably do best to bluff when receivers devalue signals. In summary, a signaller faces a situation rather like the Prisoners' Dilemma: $R > P > T > S$.

For the receiver, clearly $P' > S'$: a trusting receiver loses when an opponent is deceitful. Also, $P' > R'$, provided there is some advantage in recognising a dangerous opponent for what it is. Reacting to a large opponent as if it were small might well incur some risk. This relation deviates from the Prisoners' Dilemma. Furthermore, $P' = T'$: a receiver does not suffer a disadvantage from escalation of a signal by bluffing and devaluation. Finally, the relationship of S' and R' depends on the balance of different disadvantages. In summary, a receiver faces a situation quite different from the Prisoners' Dilemma: $P' = T' > S'$ and R'.

Nevertheless, the outcome of this game might not differ substantially from that of the Prisoners' Dilemma. In a population of trusting receivers, it clearly pays for signallers to become bluffers ($R > P$). Yet when bluffers appear, it pays for receivers to become devaluers ($T' > S'$). If $T > S$, the population would reach an evolutionarily stable state with all signallers deceitful and all receivers devaluers. Like the Prisoners' Dilemma model, the defecting strategies are not open to invasion by the alternatives.

When two individuals play this game of communication repeatedly, it is possible that 'tit-for-tat' could prevail, as it does in the Prisoners' Dilemma. Note that the receiver can do no better than P', the pay-off when honesty and trust prevail. Thus there is no incentive for the receiver to defect. If a signaller tries deceit, the receiver can switch to devaluing for the next interaction, so the signaller does worse than when honesty and trust prevail ($P > S$ and T). Thus tit-for-tat in responding might make simple honesty in signalling an evolutionarily stable strategy.

What purpose can such abstract discussion serve in understanding actual communication among animals? First, it identifies some sufficient conditions for simple honesty in signalling and trust in receiving: there must be repeated interactions over an indeterminate period between acquainted individuals. Secondly, it emphasises that the evolution of signallers' and receivers' behaviour depends on a large number of parameters. There are eight pay-offs (changes in fitness) for the four kinds of interaction. The relationships among the pay-offs in the above example for a game of communication might not apply to other possibilities for deceit, such as mimicry countered by increased discrimination by receivers. In addition, the relative frequency of each strategy in a population influences the evolution of signalling and receiving when each individual interacts with many others. None of the examples that have been suggested as involving deceitful communication has considered all the parameters necessary for a full evolutionary analysis.

5.6.4   Bluffing and ritualisation

Simple honesty, as we have just seen, is favoured between acquainted individuals that interact repeatedly. Previously we saw that unritualised signals are likewise favoured between acquainted individuals, particularly at close range. In contrast, ritualisation and escalation by bluffing and devaluation should both evolve more often for interactions between unacquainted individuals.

Thus, the same conditions that favour trust and simple honesty in signalling also favour lack of ritualisation.

When we recall that a cautious receiver creates noise in communication, an association of bluffing and ritualisation in signalling on the one hand, and of simple honesty and lack of ritualisation on the other, makes good sense. Bluffers faced with cautious receivers need to use every trick to elicit a response, including redundancy, conspicuousness, and typical intensity. Ritualised signals are just the sort that bluffers should use.

5.7   Rules in communication

A theme that runs through all sections of this chapter is the importance of rules in communication. Rules for encoding and decoding have a central place in communication for the basic reason that signals do not provide the power to produce responses directly. Every signal influences the behaviour of a receiver as a result of neuronal decoding mechanisms in the receiver that associate the stimulus of a signal with responses or changes in internal state. The encoding of a signal similarly results from mechanisms in the sender that associate internal states with actions or external changes that produce signals.

These mechanisms in the sender for encoding signals and in the receiver for decoding them do not evolve independently. Regardless of whether communication is mutualistic or manipulative, the consequences of a communicative interaction are determined by the relationship between encoding and decoding. The consequences for a sender of producing a particular signal in a particular situation depend on the receiver's mechanism for decoding that signal into responses. Likewise, the consequences for a receiver of responding to a signal in a particular way in a particular situation depend on the sender's mechanism for encoding that signal. Deceitful signalling, we have seen, takes advantage of the receiver's decoding mechanisms, and eavesdropping by receivers exploits the sender's encoding mechanisms. This interdependence of encoding and decoding mechanisms in determining the consequences of any communicatory interaction gives the evolution of communication its peculiar complexity.

This interdependence also justifies our recognition of rules in communication. Rules are simply descriptions of the coordination of prevailing mechanisms for encoding and decoding. Philo-sophers of human language in this century have placed much emphasis on rules of usage. Some, like Ludwig Wittgenstein (see Fann 1969), have emphasised that rules are not a private invention of any one mind. Instead, they result from the coordinated usage of language by a community of people. These rules of usage often cannot be explicitly formulated within the language itself; indeed, it is even difficult to enumerate all rules for the usage of a particular expression.

Coordination of encoding and decoding results in rules of usage for any system of communication. Our analysis of animal communication has shown that rules are just as central here as in human language. Furthermore, rules are as essential for mani-pulation of receivers by deceit as they are for mutualistic com-munication, since deception results from infrequent violations of established rules.

Our conclusions about evolution, however, leave open the question of the mechanisms for the development of encoding or decoding in an individual's lifetime. In fact, remarkably little is known about the ontogeny of usage, either encoding or decoding, in animal communication (for an example, see Seyfarth & Cheney 1981). Philosophers of language rightly presume that the usage of language by humans develops predominantly by learning. When they apply the term 'conventional' to rules of usage, they suggest both that rules are arbitrary, in lacking any necessary form, and that they are acquired by learning within a community of users.

For nonhuman animals, the entire question of the interactions of genes and experience in the development of rules of usage has hardly been addressed. Although we have seen how the encoding and decoding of signals evolve in synchrony, as a result of the central importance of rules in any system of communication, it is important to realise that this coordination might arise in very different ways in the ontogenetic development of individuals.

5.8   Selected reading

A very thorough review of evolutionary and other aspects of communication is provided by Green and Marler (1979). Parker (1974) develops the idea that, during aggressive interactions, animals use displays to assess the fighting ability of their opponents and use this information in deciding whether to terminate or continue a dispute. Maynard Smith (1976b) provides a concise account of the application of Games Theory to animal behaviour. He argues that, in interactions between two animals, there will usually be some conflict of interest and that what is best for one animal to do will depend on what the other does. These ideas are discussed by Caryl (1979) who suggests that they depart from a traditional ethological view that communication evolves for the mutual benefit of signaller and receiver. In a reply to Caryl, Hinde (1981) argues both that Caryl misrepresents many early ethological interpretations of communication, and that there is currently too much emphasis on the idea that communication is a means by which one animal may exploit or manipulate another, a view developed by Dawkins and Krebs (1978). The evolution of co-operative behaviour on the basis of reciprocity is discussed by Axelrod and Hamilton (1981), who analyse the Prisoners' Dilemma game outlined in this chapter. A critical discussion of the various ways that the word 'information' has been used is provided by Wiley and Richards (1983).

REFERENCES

The section(s) in which each reference appears is given after the reference

Adler J. (1976) The sensing of chemicals by bacteria. *Scientific American* **234** (4), 40–47.
4.6.1

Ali M.A. (ed.) (1978) *Sensory Ecology*. NATO Advanced Studies Institute, Series A, Life Sciences Vol. 18. Plenum Press, New York.
4.8

Altmann S.A. (1965) Sociobiology of rhesus monkeys. II. Stochastics of social communication. *Journal of Theoretical Biology* **8**, 490–522.
1.3.2

Andersson M. (1974) Temporal graphical analysis of behaviour sequences. *Behaviour* **51**, 38–48.
1.3.1

Andersson M. (1980) Why are there so many threat displays? *Journal of Theoretical Biology* **86**, 773–781.
5.4.1

Arnold S.J. (1976) Sexual behaviour, sexual interference and sexual defence in the salamanders *Ambystoma maculatum, Ambystoma tigrinum* and *Plethodon jordani*. *Zeitschrift für Tierpsychologie* **42**, 247–300.
2.8.3

Arnold S.J. (1977) The evolution of courtship behaviour in New World salamanders with some comments on Old World salamanders. In: *The Reproductive Biology of Amphibians* (ed. D.H. Taylor & S.I. Guttman), pp. 141–183. Plenum Press, New York.
2.8.3

Attneave F. (1959) *Applications of Information Theory to Psychology*. Holt, Reinhart & Winston, New York.
2.11, 2.13

Autrum H. (1940) Über Lautäusserungen und Schallwahrnehmung bei Arthrophoden. II. Das Richtungshören von *Locusta* und Versuch einer Hörtheorie für Typanalorgane von Locudistentyp. *Zeitschrift für vergleichende Physiologie* **28**, 326–352.
4.6.2

Axelrod R. & Hamilton W.D. (1981) The evolution of cooperation. *Science* **211**, 1390–1396.
5.6.2, 5.8

Baerends G.P. (1975) An evaluation of the conflict hypothesis as an explanatory principle for the evolution of displays. In: *Function and Evolution of Behaviour* (ed. G.P. Baerends, C. Beer & A. Manning), pp. 187–228. Oxford University Press, Oxford.
1.3.1, 2.8.4

190

Baerends G.P., Brouwer R. & Waterbolk H.T. (1955) Ethological studies on *Lebistes reticulatus* (Peters): I. An analysis of the male courtship pattern. *Behaviour* **8**, 249–334.
2.8.3

Bailey W.J. & Roberts J.D. (1981) The bioacoustics of the burrowing frog *Heleioporus* (Leptodactylidae). *Journal of Natural History* **15**, 693–702.
3.4.2

Bardach J.E. & Todd J.A. (1970) Chemical communication in fish. In: *Advances in Chemoreception*, Vol. 1 (ed. J.W. Johnston, D.G. Moulton & A. Turk), pp. 205–240. Appleton-Century-Crofts, New York.
1.4.2

Barlow G.W. (1977) Modal action patterns. In: *How Animals Communicate* (ed. T.A. Sebeok), pp. 98–134. Indiana University Press, Bloomington.
1.3.1, 2.2

Bastian J. (1976) Frequency response characteristics of electroreceptors in weakly electric fish (Gymnotoidei) with a pulse discharge. *Journal of Comparative Physiology* **112**, 165–190.
4.3.1

Bateson P.P.G. (1966) The characteristics and context of imprinting. *Biological Reviews* **41**, 177–220.
2.5

Bateson P.P.G. (1973) Internal influences on early learning in birds. In: *Constraints on Learning* (ed. R.A. Hinde & J. Stevenson-Hinde), pp. 101–116. Academic Press, London.
2.5

Bateson P.P.G. (1980) Optimal outbreeding and the development of sexual preferences in Japanese quail. *Zeitschrift für Tierpsychologie* **53**, 231–244.
2.6

Bateson P.P.G. (1982) Preferences for cousins in Japanese quail. *Nature* (London) **295**, 236–237.
2.6

Bateson P.P.G. (ed.) (1983) *Mate Choice.* Cambridge University Press, Cambridge.
2.8.3

Baylis J.R. (1976) A quantitative study of long term courtship. II. A comparative study of the dynamics of courtship in two New World cichlid fishes. *Behaviour* **59**, 117–161.
1.3.1, 1.3.2

Beer C.G. (1970) Individual recognition of voice in the social behavior of birds. *Advances in the Study of Behavior* **3**, 27–74.
2.5

Bell C.C. (1979) Central nervous system physiology of electroreception, a review. *Journal de Physiologie* **75**, 361–379.
4.2.3

Bell W.J. & Tobin T.R. (1982) Chemo-orientation. *Biological Reviews* **57**, 219–260.
4.6.1

Bennett M.V.L. (1971) Electroreception. In: *Fish Physiology* (ed. W.S. Hoar & D.J. Randall), pp. 493–574. Academic Press, New York.
4.2.3

Bennet-Clark H.C. (1970) The mechanism and efficiency of sound production in mole crickets. *Journal of Experimental Biology* **52**, 619–652.
3.4.2

Bennet-Clark H.C. (1971) Acoustics of insect song. *Nature* (London) **234**, 255–259.
    3.4.1, 4.3.1

Bennet-Clark H.C. & Ewing A.W. (1967) Stimuli provided by courtship of male *Drosophila melanogaster*. *Nature* (London) **215**, 669–671.
    1.4.1, 4.3.1

Bennet-Clark H.C. & Ewing A.W. (1969) Pulse interval as a critical parameter in the courtship *Drosophila melanogaster*. *Animal Behaviour* **17**, 755–759.
    1.4.1, 4.3.1

Bercken J.H.L. van den & Cools A.R. (1980) Information-statistical analysis of social interaction and communication: an analysis-of-variance approach. *Animal Behaviour* **28**, 172–188.
    1.3.2

Bertram B. (1970) The vocal repertoire of the Indian hill mynah, *Gracula religiosa*. *Animal Behaviour Monographs* **3**, 79–192.
    2.4

Birke L.I.A. (1974) Social facilitation in the Bengalese finch. *Behaviour* **48**, 111–122.
    1.1

Blum M.S. (1974) Pheromonal bases of social manifestations in insects. In: *Pheromones* (ed. M.C. Birch), pp. 190–199. North Holland, Amsterdam.
    3.2.1

Boer B.A.de (1980) A causal analysis of the territorial and courtship behaviour of *Chromis cyanea* (Pomacentridae, Pisces). *Behaviour* **73**, 1–50.
    1.3.1

Bonke D., Scheich H. & Langner G. (1979) Responsiveness of units in the auditory neostriatum of the guinea fowl *Numida meleagris* to species specific calls and synthetic stimuli. *Journal of Comparative Physiology* **132**, 243–255.
    4.5.2

Bossert W.H. (1968) Temporal patterning in olfactory communication. *Journal of Theoretical Biology* **18**, 157–170.
    3.2.2, 4.6.1

Bossert W.H. & Wilson E.O. 1963. The analysis of olfactory communication among animals. *Journal of Theoretical Biology* **5**, 443–469.
    3.2.1, 3.2.2

Bracewell R.N. (1978) *The Fourier Transform and its Applications*. McGraw-Hill, New York.
    4.3.1, 4.8

Brines M.L. & Gould J.L. (1979) Bees have rules. *Science* **206**, 571–573.
    2.9.2

Brockway B.F. (1965) Stimulation of ovarian development and egg-laying by male courtship vocalization in budgerigars (*Melopsittacus undulatus*). *Animal Behaviour* **13**, 575-578.
    1.4.1

Brooke M. de L. (1978) Sexual differences in the voice and individual vocal recognition in the Manx shearwater (*Puffinus puffinus*). *Animal Behaviour* **26**, 622–629.
    2.5

Brooks R.J. & Falls J.B. (1975a) Individual recognition by song in white-throated sparrows. I. Discrimination of songs of neighbors and strangers. *Canadian Journal of Zoology* **53**, 879–888.
    2.4, 2.5

Brooks R.J. & Falls J.B. (1975b) Individual recognition by song in white-throated sparrows. III. Song features used in individual recognition. *Canadian Journal of Zoology* **53,** 1749–1761.
1.4.1, 2.5

Brown C.H. (1982) Ventriloquial and locatable vocalizations in birds. *Zeitschrift für Tierpsychologie* **59,** 338–350.

Brown J.L. (1964) The integration of agonistic behavior in the Steller's jay *Cyanocitta stelleri* (Gmelin). *University of California Publications in Zoology* **60,** 223–328.
1.3.1

Brown R.E. (1979) Mammalian social odors, a critical review. *Advances in the Study of Behavior* **10,** 103–162.
2.4, 2.8.3

Buchler T.L., Ryan M.J. & Bartholomew G.A. (1982) Oxygen consumption during resting, calling, and nest building in the frog *Physalaemus pustulosus*. *Physiological Zoology* **55,** 10–22.
3.4

Bullock T.H. (1981) Coding and integration in receptors and central afferent systems. In: *Sense Organs* (ed. M.S. Laverack & D.J. Cosens), pp. 366–380. Blackie, Glasgow.
4.4.1

Burghardt G.M. (1970) Defining communication. In: *Advances in Chemoreception,* Vol. 1 (ed. J.W. Johnston, D.G. Moulton & A. Turk), pp. 5–18. Appleton-Century-Crofts, New York.
1.6

Busnel R.-G. & Fish J.F. (eds) (1980) *Animal Sonar Systems.* NATO Advanced Studies Institute, Series A, Life Sciences Vol. 28. Plenum Press, New York.
4.8

Cade W. (1979) The evolution of alternative male reproductive strategies in field crickets. In: *Sexual Selection and Reproductive Competition in Insects* (ed. M. Blum & N. Blum), pp. 343–380. Academic Press, New York.
3.6.2

Capranica R.R. (1965) The evoked vocal response of the bullfrog. A study of communication by sound. *Research Monographs* **33.** MIT Press, Cambridge, Mass.
4.3.1, 4.5.1

Capranica R.R. (1966) Vocal response of the bullfrog to natural and synthetic mating calls. *Journal of the Acoustical Society of America* **40,** 1131–1139.
1.4.1

Capranica R.R. (1976) Morphology and physiology of the auditory system. In: *Frog Neurobiology* (ed. R. Llinas & W. Precht), pp. 551–575. Springer-Verlag, Berlin.
4.3.1, 4.5.1

Capranica R.R. & Moffat A. (1975) Selectivity of the peripheral auditory system of spadefoot-toads (*Scaphiopus couchi*) for sounds of biological significance. *Journal of Comparative Physiology* **100,** 231–249.
4.3.1

Capranica R.R., Frischkopf L.S. & Nevo E. (1973) Encoding of geographic dialects in the auditory system of the cricket frog. *Science* **182,** 1272–1275.
3.5, 4.3.1

Carpenter C.C. (1978) Ritualistic social displays in lizards. In: *Behavior and Neurology of Lizards* (ed. N. Greenberg & P.D. Maclean), pp. 253–267. National Institute of Mental Health.
2.8.1

Carson H.L. (1982) Evolution of *Drosophila* on the newer Hawaiian volcanoes. *Heredity* **48**, 3–25.
2.3

Caryl P. (1979) Communication by agonistic displays: what can games theory contribute to ethology? *Behaviour* **68**, 136–169.
5.5.1, 5.8

Catchpole C.K. (1973) The functions of advertising song in the sedge warbler (*Acrocephalus schoenobaenus*) and the reed warbler (*A. scirpaceus*). *Behaviour* **46**, 300–320.
1.4.1

Catchpole C.K. (1979) *Vocal Communication in Birds*. Edward Arnold, London.
1.2.1

Chalmers N. (1979) *Social Behaviour in Primates*. Edward Arnold, London.
2.8.3

Charnov E.L. & Krebs J.R. (1975) The evolution of alarm calls: altruism or manipulation? *American Naturalist* **109**, 107–112.
2.10

Chatfield C. & Lemon R.E. (1970) Analysing sequences of behavioural events. *Journal of Theoretical Biology* **29**, 427–445.
1.3.1

Cherry C. (1966) *On Human Communication*, 2nd edn. MIT Press, Cambridge, Mass.
4.3.1, 5.2.1

Clayton D.A. (1978) Socially facilitated behavior. *Quarterly Review of Biology* **53**, 373–392.
1.1

Clifton P.C. (1979) The synchronization of feeding in domestic chicks by sound alone. *Animal Behaviour* **27**, 829–832.
1.1

Clutton-Brock T.H. & Albon S.D. (1979) The roaring of red deer and the evolution of honest advertisement. *Behaviour* **69**, 145–169.
2.7

Clutton-Brock T.H. & Harvey P.H. (1977) Primate ecology and social organisation. *Journal of Zoology*, London **183**, 1–39.
2.8.3

Conner W.E., Eisner T., Vander Meer R.K., Guerrero A., Ghiringelli D. & Meinwald J. (1980) Sex attractant of an arctiid moth (*Uretheisa ornatrix*): a pulsed chemical signal. *Behavioral Ecology and Sociobiology* **7**, 55–63.
3.2.2

Cox C.R. & Le Boeuf B.J. (1977) Female incitation of male competition: a mechanism of mate selection. *American Naturalist* **111**, 317–335.
5.4.1

Craik K.J.W. (1944) White plumage of seabirds. *Nature* (London) **153**, 288.
1.1

Cullen J.M. (1966) Reduction of ambiguity through ritualization. *Philosophical Transactions of the Royal Society* B **251**, 363–374.
5.2.2

Cullen J.M. (1972) Some principles of animal communication. In: *Non-verbal Communication* (ed. R.A. Hinde), pp. 101–122. Cambridge University Press, Cambridge.
1.3.2, 1.6, 2.13

Darwin C. (1871) *The Descent of Man, and Selection in Relation to Sex.* John Murray, London.
5.2.2

Darwin C. (1872) *The Expression of the Emotions in Man and Animals.* John Murray, London.
Introduction, 1.2.2, 2.8.2, 5.2.2

Davies N.B. (1978) Territorial defence in the speckled wood butterfly (*Pararge aegeria*); the resident always wins. *Animal Behaviour* **26**, 138–147.
5.5.3

Davies N.B. (1981) Calling as an ownership convention on pied wagtail territories. *Animal Behaviour* **29**, 529–534.
2.8.2

Davies N.B. & Halliday T.R. (1978) Deep croaks and fighting assessment in toads *Bufo bufo. Nature* (London) **274**, 683–685.
2.7, 5.4.2

Davies N.B. & Halliday T.R. (1979) Competitive mate searching in male common toads, *Bufo bufo. Animal Behaviour* **27**, 1253–1267.
2.7

Davis J.W.F. & O'Donald P. (1976) Sexual selection for a handicap, a critical analysis of Zahavi's model. *Journal of Theoretical Biology* **57**, 345–354.
5.4.2

Dawkins R. (1976a) Hierarchical organisation: a candidate principle for ethology. In: *Growing Points in Ethology* (ed. P.P.G. Bateson & R.A. Hinde), pp. 7–54. Cambridge University Press, Cambridge.
1.3.1

Dawkins R. (1976b) *The Selfish Gene.* Oxford University Press, Oxford.
5.4.1

Dawkins R. & Krebs J.R. (1978) Animal signals: information or manipulation? In: *Behavioural Ecology* (ed. J.R. Krebs & N.B. Davies), pp. 282–309. Blackwell Scientific Publications, Oxford.
1.1, 5.1, 5.3.1, 5.8

Dawkins R. & Krebs J.R. (1979) Arms races between and within species. *Proceedings of the Royal Society* B **205**, 489–511.
5.4.1

Dingle H. (1972) Aggressive behavior in stomatopods and the use of information theory in the analysis of animal communication. In: *Behavior of Marine Animals,* Vol. 1 (ed. H.E. Winn & B. Olla), pp. 126–156. Plenum Press, New York.
1.3.2, 1.6, 2.11

Dominey W.J. (1981) Maintenance of female mimicry as a reproductive strategy in bluegill sunfish (*Lepomis macrochirus*). *Environmental Biology of Fish* **6**, 59–64.
2.8.3

Dooling R.J., Mulligan J.A. & Miller J.D. (1971) Auditory sensitivity and song spectrum of the common canary (*Serinus canarius*). *Journal of the Acoustical Society of America* **50**, 700–709.
4.3.1

Eberhard W.G. (1977) Aggressive chemical mimicry by a bolas spider. *Science* **198**, 1173–1175.
3.6.2

Ehret G. & Gerhardt H.C. (1980) Auditory masking and effects of noise on responses of the green treefrog (*Hyla cinerea*) to synthetic mating calls. *Journal of Comparative Physiology* **141**, 13–18.
3.4.2

Embleton T.F.W., Piercy J.E. & Olson N. (1976) Outdoor sound propagation over ground of finite impedance. *Journal of the Acoustical Society of America* **59**, 267–277.
3.4.1

Emlen S.T. (1972) An experimental analysis of the parameters of bird song eliciting species recognition. *Behaviour* **41**, 130–171.
1.4.1, 2.3, 4.5.2

Erickson R.P. (1974) Parallel population neural coding in feature extraction. In: *Neurosciences Third Study Program* (ed. F.O. Schmitt & F.G. Worden), pp. 155–169. MIT Press, Cambridge, Mass.
4.4.2

Ewert J.P., Capranica R.R. & Ingle D.J. (in press) *Advances in Vertebrate Neuroethology*. NATO Advanced Science Institute. Plenum Press, London.
4.1, 4.5.2

Ewing A.W. & Bennet-Clark H.C. (1968) The courtship songs of *Drosophila*. *Behaviour* **31**, 288–301.
1.4.1

Fagen R.M. & Mankovich N.J. (1980) Two-act transitions, partitioned contingency tables, and the 'significant' cells problem. *Animal Behaviour* **28**, 1017–1023.
1.3.1

Falls J.B. (1963) Properties of bird song eliciting responses from territorial males. *Proceedings of the 13th International Ornithological Congress*, pp. 259–271. American Ornithologists' Union.
4.5.2

Falls J.B. & Brooks R.J. (1975) Individual recognition by song in white-throated sparrows. II. Effects of location. *Canadian Journal of Zoology* **53**, 1412–1420.
1.4.1, 4.6.1, 5.6.2

Fann K.T. (1969) *Wittgenstein's Conception of Philosophy*. University of California Press, Berkeley.
5.7

Farkas S.R. & Shorey H.H. (1972) Chemical trail-following by flying insects: a mechanism for orienting to a distant odor source. *Science* **178**, 67–68.
3.2.2

Farkas S.R. & Shorey H.H. (1974) Mechanisms of orientation to a distant pheromone source. In: *Pheromones* (ed. M.C. Birch), pp. 81–95. North Holland, Amsterdam.
3.2.1, 3.2.2

Fay R.R. (1981) Coding of acoustic information in the eighth nerve. In: *Hearing and Sound Communication in Fishes* (ed. W.N. Tavolga, A.N. Popper & R.R. Fay), pp. 189–221. Springer-Verlag, Berlin.
4.4.2

Feng A.S., Narins P.M. & Capranica R.R. (1975) Three populations of primary auditory fibers in the bullfrog (*Rana catesbeiana*): their peripheral origins and

frequency sensitivities. *Journal of Comparative Physiology* **100**, 221–229.
4.2.2

Fish J.F. & Offutt G.C. (1972) Hearing thresholds from toad fish, *Opsanus tau,*
measured in the laboratory and field. *Journal of the Acoustical Society of America*
51, 1318–1321.
4.3.1

Fraenkel G.S. & Gunn D.L. (1940) *The Orientation of Animals.* Oxford University
Press, Oxford.
4.6.1

Franceschini N., Hardie R., Ribi W. & Kirschfeld K. (1981) Sexual dimorphism in a
photoreceptor. *Nature* (London) **291**, 241–244.
4.2.4

Frisch K. von (1976) *The Dance Language and Orientation of Bees.* Harvard University
Press, Cambridge, Mass.
2.9.2

Frischkopf L.S. & Goldstein M.H. (1963) Responses to acoustic stimuli from single
units in the eighth nerve of the bullfrog. *Journal of the Acoustical Society of
America* **35**, 1219–1228.
4.3.1

Frischkopf L.S., Capranica R.R. & Goldstein M.H. (1968) Neural coding in the
bullfrog's auditory system – a teleological approach. *Proceedings I.E.E.E.* **56**,
969–980.
4.3.1

Galusha J.G. & Stout J.F. (1977) Aggressive communication by *Larus glaucescens.*
Part IV. Experiments on visual communication. *Behaviour* **62**, 222–235.
1.4.3

Ganyard M.C. Jr & Brady U.E. (1972) Interspecific attraction in Lepidoptera in the
field. *Annals of the Entomological Society of America* **65**, 1279–1282.
3.2.3

Gerhardt H.C. (1974) The significance of some spectral features in mating call
recognition in the green treefrog (*Hyla cinerea*). *Journal of Experimental Biology*
61, 229–241.
1.4.1

Gerhardt H.C. (1976) Significance of two frequency bands in long distance vocal
communication in the green treefrog. *Nature* (London) **261**, 692–694.
3.4.2

Gerhardt H.C. (1978a) Temperature coupling in the vocal communication system of
the gray treefrog, *Hyla versicolor. Science* **199**, 992–994.
3.6.1

Gerhardt H.C. (1978b) Discrimination of intermediate sounds in a synthetic call
continuum by female green treefrogs. *Science* **199**, 1089–1091.
3.4.2

Gerhardt H.C. (1982) Sound pattern recognition in some North American treefrogs
(Anura: Hylidae): implications for male choice. *American Zoologist* **22**, 581–595.
3.6.1

Gerhardt H.C. & Mudry K.M. (1980) Temperature effects on frequency preferences
and mating call frequencies in the green treefrog, *Hyla cinerea. Journal of
Comparative Physiology* **137**, 1–6.
3.6

Goldman P. (1973) Song recognition in field sparrows. *Auk* **90**, 106–113.
1.4.1

Goodman L.A. (1968) The analysis of cross-classified data: independence, quasi-independence and interactions in contingency tables with or without missing entries. *Journal of the American Statistical Association* **63**, 1091–1131.
1.3.1

Gorman M.L. (1976) A mechanism for individual recognition by odour in *Herpestes auropunctatus* (Carnivora: Viverridae). *Animal Behaviour* **24**, 141–145.
2.5

Gould J.L. (1976) The dance-language controversy. *Quarterly Review of Biology* **51**, 211–244.
2.9.2

Green S. & Marler P. (1979) The analysis of animal communication. In: *Handbook of Behavioral Neurobiology*, Vol. 3 (ed. P. Marler & J.G. Vandenbergh), pp. 73–158. Plenum Press, New York.
1.1, 1.6, 5.2.1, 5.8

Greenberg L. (1979) Genetic component of bee odor in kin recognition. *Science* **206**, 1095–1097.
2.6

Greig-Smith P.W. (1982) Song rates and parental care by individual male stonechats (*Saxicola torquata*). *Animal Behaviour* **30**, 245–252.
2.8.3

Grinnell A.D. & Schnitzler H.U. (1977) Directional sensitivity of echolocation in the horseshoe bat *Rhinolophus ferrume quinum*. *Journal of Comparative Physiology* **116**, 63–76.
4.6.1

Guthrie D.M. (1980) *Neuroethology: An Introduction*. Blackwell Scientific Publications, Oxford.
4.1

Halliday T.R. (1975) An observational and experimental study of sexual behaviour in the smooth newt, *Triturus vulgaris* (Amphibia; Salamandridae). *Animal Behaviour* **23**, 291–322.
1.4.3

Halliday T.R. (1976) The libidinous newt. An analysis of variations in the sexual behaviour of the male smooth newt, *Triturus vulgaris*. *Animal Behaviour* **24**, 398–414.
2.8.3

Halliday T.R. (1977) The courtship of European newts. An evolutionary perspective. In: *The Reproductive Biology of Amphibians* (ed. D.H. Taylor & S.I. Guttman), pp. 185–232. Plenum Press, New York.
2.8.3

Halliday T.R. (1978) Sexual selection and mate choice. In: *Behavioural Ecology. An Evolutionary Approach* (ed. J.R. Krebs & N.B. Davies), pp. 180–213. Blackwell Scientific Publications, Oxford.
2.3.1, 2.8.3

Halliday T.R. (1983) The study of mate choice. In: *Mate Choice* (ed. P.P.G. Bateson), pp. 3–32. Cambridge University Press, Cambridge.
2.8.3

Halliday T.R. & Houston A.I. (1978) The newt as an honest salesman. *Animal Behaviour* **26**, 1273–1274.
2.8.3

Hamilton W.D. (1964) The genetical evolution of social behaviour, I and II. *Journal of Theoretical Biology* **7**, 1–52.
5.3.1

Harris M.A. & Lemon R.E. (1974) Songs of song sparrows: reactions of males to songs of different localities. *Condor* **76**, 33–44.
1.4.1

Harvey P.H. & Greenwood P.J. (1978) Anti-predator defence strategies: some evolutionary problems. In: *Behavioural Ecology* (ed. J.R. Krebs & N.B. Davies), pp. 129–151. Blackwell Scientific Publications, Oxford.
1.1, 2.10

Hausen K. & Strausfield N.J. (1980) Sexually dimorphic interneuron arrangements in the fly visual system. *Proceedings of the Royal Society of London* B **208**, 57–71.
4.2.4

Hawkins A.D. (1981) The learning abilities of fish. In: *Hearing and Sound Communication in Fishes* (ed. W.N. Tavolga, A.N. Popper & R.R. Fay), pp. 109–133. Springer-Verlag, Berlin.
4.3.1

Hawkins A.D. & Sand D. (1977) Directional hearing in the median vertical plane by the cod. *Journal of Comparative Physiology* **122**, 1–8.
4.6.2

Hayward J.L. Jr, Gillett H. & Stout J. (1977) Aggressive communication in *Larus glaucescens*. Part V. Orientation and sequences of behaviour. *Behaviour* **62**, 236–276.
1.4.3

Hazlett B.A. (1972) Stereotypy of agonistic movements in the spider crab *Microphrys bicornutus*. *Behaviour* **42**, 270–278.
2.2

Hazlett B.A. (1978) Shell exchanges in hermit crabs: aggression, negotiation, or both? *Animal Behaviour* **26**, 1278–1279.
2.10.2

Hazlett B.A. (1983) Interspecific negotiations: mutual gain in exchanges of a limiting resource. *Animal Behaviour* **31**, 160–163.
2.10.2

Hazlett B.A. & Bossert W.H. (1965) A statistical analysis of the aggressive communications systems of some hermit crabs. *Animal Behaviour* **13**, 357–373.
2.11

Heiligenberg W. (1977) Principles of electrolocation and jamming avoidance in electric fish. A neuroethological approach. In: *Studies in Brain Function I*, pp. 1–85. Springer-Verlag, Berlin.
4.6.1

Heiligenberg W. & Altes R. (1978) Phase sensitivity in electroreception. *Science* **199**, 1001–1004.
4.3.2

Henwood K. & Fabrick A. (1979) A quantitative analysis of the dawn chorus: temporal selection for communicatory optimization. *American Naturalist* **114**, 260–274.
3.4.2

Hill K.G. (1974) Carrier frequency as a factor in phonotactic behavior of female crickets (*Teleogryllus commodus*). *Journal of Comparative Physiology* **93**, 7–18.
4.3.1

Hill K.G. & Boyan G.S. (1976) Directional hearing in crickets. *Nature* (London) **262**, 390–391.
4.6.2

Hinde R.A. (1953) The conflict between drives in the courtship and copulation of the chaffinch. *Behaviour* **5**, 1–31.
2.8.4

Hinde R.A. (1970) *Animal Behaviour*, 2nd edn. McGraw-Hill, New York.
1.3.1

Hinde R.A. (ed.) (1972) *Non-verbal Communication*. Cambridge University Press, Cambridge.
2.13

Hinde R.A. (1975) *Biological Bases of Human Social Behaviour*. McGraw-Hill, New York.
1.6

Hinde R.A. (1981) Animal signals: ethological and games-theory approaches are not incompatible. *Animal Behaviour* **29**, 535–542.
1.1, 5.5.1, 5.8

Hölldobler B. (1971) Communication between ants and their guests. *Scientific American* **224**, 86–93.
2.10.3

Hopkins C.D. (1974) Electric communication in fish. *American Scientist* **62**, 426–437.
4.2.3

Hopkins C.D. (1976) Stimulus filtering and electroreception: tuberous electroreceptors in three species of gymnotoid fish. *Journal of Comparative Physiology* **111**, 171–207.
4.3.1

Hopkins C.D. (1977) Electric communication. In: *How Animals Communicate* (ed. T.A. Sebeok), pp. 263–289. Indiana University Press, Bloomington.
4.2.3, 4.3.1

Hopkins C.D. (1981) On the diversity of electric signals in a community of mormyrid electric fish in West Africa. *American Zoologist* **21**, 211–222.
4.2.3

Hopkins C.D. & Bass A.H. (1981) Temporal coding of species recognition signals in an electric fish. *Science* **212**, 85–87.
4.3.2, 4.4.2

Hopkins C.D. & Heiligenberg W. (1978) Evolutionary designs for electric signals and electroreceptors in Gymnotoid fishes of Surinam. *Behavioral Ecology and Sociobiology* **3**, 113–134.
4.3.1

Hover E.L. & Jenssen T.A. (1976) Descriptive analysis and social correlates of agonistic displays of *Anolis limifrons* (Sauria, Iguanidae). *Behaviour* **58**, 173–191.
2.8.1

Howard R.D. (1974) The influence of sexual selection and interspecific competition on mockingbird song (*Mimus polyglottos*). *Evolution* **28**, 428–438.
2.8.3

Hubel D.H. & Wiesel T.N. (1959) Receptive fields of single neurons in the cat's striate cortex. *Journal of Physiology* **148**, 574–591.
4.5

Huber F. (1977) Lautäusserungen und Lauterkennen bei Insekten (Grillen). In: *26. Jahresfeier der Rheinisch-Westfälischen Akademie der Wissenschaften* **265**, 15–66. Westdeutscher Verlag, Opladen.
4.3.1, 4.6.2

Hughes A. (1977) The topography of vision in mammals. In: *Handbook of Sensory Physiology* Vol. II/5 (ed. F. Crescitelli), pp. 613–756. Springer-Verlag, Berlin.
3.3.5

Huxley J.S. (1914) The courtship habits of the great crested grebe (*Podiceps cristatus*); with an addition to the theory of sexual selection. *Proceedings of the Zoological Society of London* **35**, 491–562.
Introduction

Jacobson M. (1972) *Insect Sex Pheromones*. Academic Press, New York.
1.4.2

Jenssen T.A. (1971) Display analysis of *Anolis nebulosus*. *Copeia* 197–209.
5.2.2

Jenssen T.A. (1977) Evolution of Anoline lizard display behavior. *American Zoologist* **17**, 203–215.
2.8.1

Jenssen T.A. (1979) Display modifiers of *Anolis opalinus* (Lacertilia: Iguanidae). *Herpetologica* **35**, 21–30.
2.8.1

Jerlov N.G. (1976) *Marine Optics*. Elsevier, Amsterdam.
3.3.3

Johnson C. (1959) Genetic incompatibility in the call-races of *Hyla versicolor*. *Copeia* 327–335.
2.3

Jouventin P., Guillotin M. & Cornet A. (1979) Le chant du manchot empereur et sa signification adaptive. *Behaviour* **70**, 231–250.
5.6.2

Kaae R.S. & Shorey H.H. (1972) Sex pheromones of noctuid moths. XXVII. Influence of wind velocity on sex pheromone releasing behavior of *Trichoplusia* females. *Annals of the Entomological Society of America* **65**, 436–440.
3.2.1

Kaissling K.E. (1971) Insect olfaction. In: *Handbook of Sensory Physiology* Vol. IV (ed. L.M. Beidler), pp. 351–431. Springer-Verlag, Berlin.
4.2.1

Kaissling K.E. & Priesner R. (1970) Die Riechswelle des Seidenspinners. *Naturwissenschaften* **57**, 23–28.
4.2.1, 4.4.1

Kennedy J.S. & Marsh D. (1974) Pheromone-regulated anemotaxis in flying moths. *Science* **184**, 999–1001.
3.2.2

Kirk V.M. & Dupraz B.J. (1972) Discharge behavior of female ground beetle, *Pterostichus lucublandus* (Coleoptera: Carabidae), used as a defense against males. *Annals of the Entomological Society of America* **65**, 513.
2.8.3

Kirschfeld K. (1976) The resolution of lens and compound eyes. In: *Neural Principles in Vision* (ed. F. Zettler & R. Weiler), pp. 354–369. Springer-Verlag, Berlin.
4.6.1

Knudsen E.I. (1975) Spatial aspects of the electric fields generated by weakly electric fish. *Journal of Comparative Physiology* **99**, 103–118.
4.2.3

Knudsen E.I. & M. Konishi (1978) A neural map of auditory space in the owl. *Science* **200**, 795–797.
4.6.1, 4.6.2

References

Knudsen E.I. & Konishi M. (1979) Mechanisms of sound localization in the barn owl (*Tyto alba*). *Journal of Comparative Physiology* **133**, 13–21.
4.6.1, 4.6.2

Kohda M. (1981) Interspecific territoriality and agonistic behaviour of a temperate pomacentrid fish, *Eupomacentrus altus* (Pisces; Pomacentridae). *Zeitschrift für Tierpsychologie* **56**, 205–216.
2.10.1

Konishi M. (1970) Comparative neurophysiological studies of hearing and vocalizations in song birds. *Zeitschrift für vergleichende Physiologie* **66**, 257–272.
4.3.1

Konishi M. (1973) Locatable and non-locatable acoustic signals for barn owls. *American Naturalist* **107**, 775–785.
4.6.2

Krebs J.R. (1974) Colonial nesting and social feeding as strategies for exploiting food sources in the great blue heron (*Ardea herodias*). *Behaviour* **51**, 99–134.
2.9

Krebs J.R. (1976) Habituation and song repertoires in the great tit. *Behavioral Ecology and Sociobiology* **1**, 215–227.
1.4.1

Krebs J.R. (1977) The significance of song repertoires: the Beau Geste hypothesis. *Animal Behaviour* **25**, 475–478.
5.4.1

Krebs J.R. & Kroodsma D.E. (1980) Repertoires and geographical variation in bird song. *Advances in the Study of Behavior* **11**, 143–177.
5.4.1

Krebs J.R., Ashcroft R. & Webber M.I. (1978) Song repertoires and territory defence. *Nature* (London) **271**, 539–542.
1.4.1, 5.4.1

Krebs J.R., Avery M.I. & Cowie R.J. (1981) Effect of removal of mate on the singing behaviour of great tits. *Animal Behaviour* **29**, 635–637.
1.4.1

Kroodsma D.E. (1976) Reproductive development in a female song bird: differential stimulation by quality of male song. *Science,* **192**, 574–575.
1.4.1

Kroodsma D.E. (1977) Correlates of song organization among North American wrens. *American Naturalist* **111**, 995–1008.
5.2.2

Kruijt J.P. & Hogan J.A. (1967) Social behaviour on the lek in black grouse, *Lyrurus t. tetrix* L. *Ardea* **55**, 203–240.
2.8.3

Kummer H. (1971) *Primate Societies. Group Techniques of Ecological Adaptation.* Aldine Atherton, Chicago.
Introduction

Lall B., Seliger H.H., Biggley W.H. & Lloyd J.E. (1980) Ecology of colors of firefly bioluminescence. *Science* **210**, 560–562.
3.3.6

Land M.F. (1981) Optics and vision in invertebrates. In: *Handbook of Sensory Physiology* Volume VII/6B (ed. H. Autrum), pp. 471–592. Springer-Verlag, Berlin.
4.2.4, 4.6.1, 4.8

Land M.F. & Collett T. (1974) Chasing behaviour of houseflies (*Fannia canicularis*). A description and analysis. *Journal of Comparative Physiology* **89**, 331–357.
4.2.4

Le Boeuf B.J. (1974) Male–male competition and reproductive success in elephant seals. *American Zoologist* **14**, 163–176.
5.4.1

Lee Y.W. (1960) *Statistical Theory of Communication*. John Wiley & Sons, New York.
4.3.2

Lemon R.E. & Chatfield C. (1971) Organization of song in cardinals. *Animal Behaviour* **19**, 1–17.
1.3.1

Leppelsack H.J. & Vogt M. (1976) Responses of auditory neurons in the forebrain of a songbird to stimulation with species-specific sounds. *Journal of Comparative Physiology* **107**, 263–274.
4.5.2

Lettvin J.W., Maturana W.S., McCulloch W.S. & Pitts W.S. (1959) What the frog's eye tells the frog's brain. *Proceedings I.R.E.* **47**, 1940–1951.
4.2, 4.5

Lewis D.B. (1974) The physiology of the tettigoniid ear. IV. A new hypothesis for acoustic orientation behaviour. *Journal of Experimental Biology* **60**, 861–869.
4.6.2

Lewis D.B. & Coles R. (1980) Sound localization in birds. *Trends in Neurosciences* May, 102–105.
3.4.3, 4.6.2

Linsenmair K.E. (1972) Die Bedeutung familienspezifischer 'Abzeichen' für den familienzusammenhalt bei der sozialen Wüstenassel *Hemilepistus reaumuri* Audouin u. Savigny (Crustacea, Isopoda, Oniscoidea). *Zeitschrift für Tierpsychologie* **31**, 131–162.
2.6

Littlejohn M.J. (1977) Long-range acoustic communication in anurans: an integrated and evolutionary approach. In: *The Reproductive Biology of Amphibians* (ed. D.H. Taylor & S.I. Guttman), pp. 263–294. Plenum Press, New York.
2.3

Littlejohn M.J. & Martin A.A. (1969) Acoustic interaction between two species of leptodactylid frogs. *Animal Behaviour* **17**, 785–791.
3.5

Lloyd J.E. (1966) Studies on the flash communication system in *Photinus* fireflies. *Miscellaneous Publications of the Museum of Zoology, University of Michigan* No. 130.
2.3, 2.8.3

Lloyd J.E. (1971) Bioluminescent communication in insects. *Annual Review of Entomology* **16**, 97–122.
2.3, 2.8.3

Lloyd J.E. (1977) Bioluminescence and communication. In: *How Animals Communicate* (ed. T.A. Sebeok), pp. 164–183. Indiana University Press, Bloomington.
3.3.6

Lloyd J.E. (1980) Male *Photuris* fireflies mimic sexual signals of their females' prey. *Science* **210**, 669–671.
3.6.2

Lloyd J.E. (1981) Mimicry in the sexual signals of fireflies. *Scientific American* **245** July, 111–117.
3.5, 3.6.2, 3.8

Löfquist M.J. & Bergström G. (1980) Nerol-derived volatile signals as a biochemical basis for reproductive isolation between sympatric populations of three species of ant-lions (Neuroptera: Myrmeleontidae). *Insect Biochemistry* **10**, 1–10.
2.3

Loftus-Hills J.J. & Johnstone B.M. (1970) Auditory function, communication and the brain-evoked response in anuran amphibians. *Journal of the Acoustical Society of America* **47**, 1131–1138.
3.5, 4.3.1

Lorenz K. (1935) Der Kumpan in der Umwelt des Vogels. *Journal für Ornithologie* **83**, 137–213 and 289–413.
4.5

Lorenz K. (1937) Uber die Bildung des Instinktbegriffes. *Naturwissenschaften* **25**, 289–300.
2.2

Lorenz K. (1941) Vergleichende Bewegungsstudien bei Anatiden. *Journal für Ornithologie* **89**, 194–294.
Introduction

Losey G.S. Jr (1978) Information theory and communication. In: *Quantitative Ethology* (ed. P.W. Colgan), pp. 43–78. John Wiley & Sons, New York.
1.3.2

Lythgoe J. (1972) The adaptation of visual pigments to their photic environment. In: *Handbook of Sensory Physiology* Vol. VII/1 (ed. H.J.A. Dartnall), pp. 566–603. Springer-Verlag, Berlin.
4.3.1

Lythgoe J.N. (1979) *The Ecology of Vision*. Clarendon Press, Oxford.
3.3.1, 3.3.3, 3.3.5, 3.3.6, 3.8, 4.3.1, 4.8

MacKay D.M. (1972) Formal analysis of communicative processes. In: *Non-verbal Communication* (ed. R.A. Hinde), pp. 3–25. Cambridge University Press, Cambridge.
1.1, 1.6

MacNally R.C. (1981) On the reproductive energetics of chorusing males: energy depletion profiles, restoration and growth in two sympatric species of *Ranidella* (Anura). *Oecologia* **51**, 181–188.
3.4

MacNally R.C. & Young D. (1981) Song energetics of the bladder cicada, *Cystosoma saundersii. Journal of Experimental Biology* **90**, 185–196.
3.4

Marler P. (1955) Characteristics of some animal calls. *Nature* (London) **176**, 6–7.
1.2.2, 2.9.1, 4.6.2

Marler P. (1959) Developments in the study of animal communication. In: *Darwin's Biological Work* (ed. P.R. Bell), pp. 150–206. Cambridge University Press, Cambridge.
3.4.3

Marler P. (1967) Animal communication signals. *Science* **157**, 769–774.
1.1

Marler P. (1973) A comparison of vocalizations of red-tailed monkeys and blue monkeys, *Cercopithecus ascanius* and *C. mitis*, in Uganda. *Zeitschrift für*

Tierpsychologie **33**, 223–247.
5.2.2

Marler P. & Hamilton W.J. (1967) *Mechanisms of Animal Behaviour.* John Wiley & Sons, New York.
4.8

Marler P. & Mundinger P.C. (1975) Vocalizations, social organization and breeding biology of the twite *Acanthus flavirostris. Ibis* **117**, 1–17.
1.3.1, 2.5

Marler P. & Peters S. (1977) Selective vocal learning in a sparrow. *Science* **198**, 519–521.
4.5.2

Marten K. & Marler P. (1977) Sound transmission and its significance for animal vocalization. I. Temperate habitats. *Behavioral Ecology and Sociobiology* **2**, 271–290.
3.4.2

Martin D.J. (1980) Response of male Fox sparrows to broadcast of particular conspecific songs. *Wilson Bulletin* **92**, 21–32.
1.4.1

Martin H. (1964) Nahorientierung der Biene im Duftfeld, zugleich ein Nachweis für die Osmotropotaxis bei Insekten. *Zeitschrift für vergleichende Physiologie* **48**, 481–533.
4.6.1

Maurus M. & Ploog D. (1971) Social signals in squirrel monkeys: analysis by cerebral radio stimulation. *Experimental Brain Research* **12**, 171–183.
1.4.3

Maurus M. & Pruscha H. (1972) Quantitative analysis of behavioral sequences elicited by automated telestimulation in squirrel monkeys. *Experimental Brain Research* **14**, 372–394.
1.4.3

Maynard Smith J. (1965) The evolution of alarm calls. *American Naturalist* **94**, 59–63.
2.10

Maynard Smith J. (1972) *On Evolution.* Edinburgh University Press, Edinburgh.
1.3.1

Maynard Smith J. (1974) The theory of games and the evolution of animal conflicts. *Journal of Theoretical Biology* **47**, 209–221.
5.5.2, 5.5.3, 5.5.4

Maynard Smith J. (1976a) Sexual selection and the handicap principle. *Journal of Theoretical Biology* **57**, 239–242.
5.4.2

Maynard Smith J. (1976b) Evolution and the theory of games. *American Scientist* **64**, 41–45.
5.5.3, 5.5.4, 5.8

Maynard Smith J. (1979) Game theory and the evolution of behaviour. *Proceedings of the Royal Society* B **205**, 475–488.
5.5.4

Maynard Smith J. & Parker G.A. (1976) The logic of asymmetric contests. *Animal Behaviour* **24**, 159–175.
5.5.2, 5.5.4

Michael R.P. & Keverne E.B. (1970) Primate sex pheromones of vaginal origin. *Nature* (London) **225**, 84–85.
1.4.3

Michelsen A. (1978) Sound reception in different environments. In *Perspectives in Sensory Ecology* (ed. A.B. Ali). Plenum Press, New York.
3.4.1

Michelsen A. & Larsen O.N. (1978) Biophysics of the ensiferan ear. I. Tympanal vibrations in bush crickets (Tettigoniidae) studied with laser vibrometers. *Journal of Comparative Physiology* **123**, 193–203.
4.6.2

Moller P. & Szabo T. (1981) Lesions in the nucleus mesencephali exterolateralis: effects on electrocommunication in the Mormyrid fish *Gnathonemus petersii* (Mormyriformes). *Journal of Comparative Physiology* **144**, 327–333.
4.2.3

Morgan B.J.T., Simpson M.J.A., Hanby J.P. & Hall-Craggs J. (1976) Visualizing interaction and sequential data in animal behaviour: theory and application of cluster-analysis methods. *Behaviour* **56**, 1–43.
1.3.1

Morris D. (1957) 'Typical intensity' and its relation to the problem of ritualisation. *Behaviour* **11**, 1–12.
1.3.1, 2.2, 5.2.2

Morton E.S. (1975) Ecological sources of selection on avian sounds. *American Naturalist* **109**, 17–34.
3.4.2

Morton E.S. (1977) On the occurrence and significance of motivation-structural rules in some bird and mammal sounds. *American Naturalist* **111**, 855–869.
1.2.2, 2.8.2, 5.4.2

Moseley L.J. (1979) Individual auditory recognition in the least tern (*Sterna albifrons*). *Auk* **96**, 31–39.
5.2.2, 5.6.2

Mudry K.M. Constantine-Paton M. & Capranica R.R. (1977) Auditory sensitivity of the diencephalon of the leopard frog *Rana pipiens*. *Journal of Comparative Physiology* **114**, 1–13.
4.5.1

Muntz W.R.A. (1975) The visual consequences of yellow filtering pigments in the eyes of fishes occupying different habitats. In: *Light as an Ecological Factor* Vol. II (ed. G.C. Evans, R. Bainbridge & O. Rackham), pp. 271–287. Blackwell Scientific Publications, Oxford.
3.3.4

Munz F.W. & McFarland W.N. (1977) Evolutionary adaptations of fishes to the photic environment. In: *Handbook of Sensory Physiology* Vol. VII/5 (ed. F. Crescitelli), pp. 193–274. Springer-Verlag, Berlin.
3.3.2, 3.3.3

Murphy R.K. & Zaretsky M. (1972) Orientation to calling songs by female crickets, *Scapsipedus marginatus* (Gryllidae). *Journal of Experimental Biology* **56**, 335–352.
4.6.2

Murray B.G. (1971) The ecological consequences of interspecific territoriality in birds. *Ecology* **52**, 414–423.
2.10.1

Myrberg A.A. & Spires J.Y. (1980) Hearing in damsel fishes: an analysis of signal detection among closely related species. *Journal of Comparative Physiology* **140**, 135–144.
4.3.1

Myrberg A.A., Gordon C.R. & Kimley A.P. (1976) Attraction of free ranging sharks by low frequency sound, with comments on its biological significance. In: *Sound Reception in Fish* (ed. A. Schnijf & A. Hawkins), pp. 205–228. Elsevier, New York.
4.6.2

Narins P.M. & Capranica R.R. (1976) Sexual differences in the auditory system of the tree frog *Eleutherodactylus coqui. Science* **192**, 378–380.
4.2.2

Narins P.M. & Capranica R.R. (1980) Neural adaptations for processing the two-note call of the Puerto Rican treefrog, *Eleutherodactylus coqui. Brain, Behavior and Evolution* **17**, 48–66.
2.2.1

Nelson B. (1979) *The Gannet.* T. & A.D. Poyser, Berkhamsted.
1.1

Neuweiler G. (1979) Auditory processing of echoes: peripheral processing. In: *Animal Sonar Systems* (ed. R.-G. Busnel & J.E. Fish), pp. 519–548. NATO Advanced Studies Institute, Series A, Life Sciences Vol. 28. Plenum Press, New York.
4.5.2

Nice M.M. (1943) Studies in the life history of the song sparrow. II. The behavior of the song sparrow and other passerines. *Transactions of the Linnaean Society of New York* **6**, 1–324
1.4.1

Nielsen D.G. & Balderston C.P. (1973) Evidence for intergeneric sex attraction among aegeriids. *Annals of the Entomological Society of America* **66**, 227–228.
3.2.3

Nocke H. (1975) Physical and physiological properties of the tettigoniid (grasshopper) ear. *Journal of Comparative Physiology* **100**, 25–57.
4.6.2

Nuechterlein G.L. (1981) Variations and multiple functions of the advertising display of western grebes. *Behaviour* **76**, 289–317.
2.2.2

Owings D.H. & Virginia R.A. (1978) Alarm calls of California ground squirrels (*Spermophilus beecheyi*). *Zeitschrift für Tierpsychologie* **46**, 48–70.
1.2.2

Parker G.A. (1974) Assessment strategy and the evolution of fighting behaviour. *Journal of Theoretical Biology* **47**, 223–243.
5.5.4. 5.8

Parker G.A. & Rubenstein D.I. (1981) Role assessment, reserve strategy, and acquisition of information in asymmetric animal conflicts. *Animal Behaviour* **29**, 221–240.
5.5.4

Parker G.A. & Thompson E.A. (1980) Dungfly struggles: a test of the war of attrition. *Behavioral Ecology and Sociobiology* **7**, 34–44.
5.5.4

Paul R.C. & Walker T.J. (1979) Arboreal singing in a burrowing cricket, *Anurogryllus arboreus. Journal of Comparative Physiology* **132**, 217–223.
3.4.2

Payne R. (1971) Acoustic location of prey by barn owls (*Tyto alba*). *Journal of Experimental Biology* **54**, 535–573.
4.6.2

Perkel D.H. & Bullock T.H. (1968) Neural coding. *Neurosciences Research Progam Bulletin* **6**, 221–348.
4.4.1

Perrill S.A., Gerhardt H.C., Daniel R.A. (1978) Sexual parasitism in the green treefrog (*Hyla cinerea*). *Science* **200**, 1179–1180.
3.6.2

Peters S.S., Searcy W.A. & Marler P. (1980) Species song discrimination in choice experiments with territorial male swamp and song sparrows. *Animal Behaviour* **28**, 393–404.
4.5.2

Piercy J.E. & Embelton T.F.W. (1977) Review of noise propagation in the atmosphere. *Journal of the Acoustical Society of America* **61**, 1403–1418.
3.4.1

Popper A.N. & Fay R.R. (1973) Sound detection and processing by teleost fishes: a critical review. *Journal of the Acoustical Society of America* **53**, 1515–1529.
4.3.1

Popov A.V. & Shuvalov V.F. (1977) Phonotactic behavior of crickets. *Journal of Comparative Physiology* **119**, 111–126.
4.6.2

Prozesky-Schulze L., Prozesky O.P.M., Anderson F. & Merwe G.J.J. van der (1975) Use of a self-made baffle by a tree cricket. *Nature* (London) **255**, 142–143.
3.4.2

Pruscha H. & Maurus M. (1976) The communicative function of some agonistic behaviour patterns in squirrel monkeys: the relevance of social context. *Behavioral Ecology and Sociobiology* **1**, 185–214.
1.4.3

Pusey A.E. (1980) Inbreeding avoidance in chimpanzees. *Animal Behaviour* **28**, 543–552.
2.6

Raisbeck G. (1963) *Information Theory.* MIT Press, Cambridge, Mass.
5.2.2

Rapoport A. (1960) *Fights, Games and Debates.* University of Michigan Press, Ann Arbor.
5.6.2

Rapoport A. & Chammah A.M. (1965) *Prisoner's Dilemma: A Study in Conflict and Cooperation.* University of Michigan Press, Ann Arbor.
5.6.2

Rasa O.A.E. (1973) Marking behaviour and its social significance in the African dwarf mongoose, *Helogale undulata rufula. Zietschrift für Tierpsychologie* **32**, 293–318.
2.4, 2.5

Reed T.M. (1982) Interspecific territoriality in the chaffinch and the great tit on islands and the mainland of Scotland: playback and removal experiments. *Animal Behaviour* **30**, 171–181.
1.4.1, 2.10.1

Rheinlaender J. & Römer H. (1980) Bilateral coding of sound direction in the CNS of the bushcricket *Tettigonia viridissima* L. (Orthoptera, Tettigoniidae). *Journal of Comparative Physiology* **140**, 101–111.
4.6.2

Rheinlaender J., Gerhardt H.C., Yager D.D. & Capranica R.R. (1979) Accuracy of phonotaxis by the green treefrog (*Hyla cinerea*). *Journal of Comparative Physiology* **133**, 247–255.
3.4.3

Rhijn J.G. van (1980) Communication by agonistic displays: a discussion. *Behaviour* **74**, 284–293.
5.5.1

Richards D.G. (1981a) Estimation of distance of singing conspecifics by the Carolina wren. *Auk* **98**, 127–133.
1.4.1

Richards D.G. (1981b) Alerting and message components in songs of rufous-sided towhees. *Behaviour* **76**, 223–249.
5.2.2

Riechert S.E. (1978) Games spiders play: behavioral variability in territorial disputes. *Behavioral Ecology and Sociobiology* **3**, 135–162.
5.5.2

Roelofs W. (1979) Production and perception of Lepidopterous pheromone blends. In: *Chemical Ecology: Odour Communication in Animals* (ed. F.J. Ritter), pp. 159–168. Elsevier/North Holland, Amsterdam.
3.2.3

Rohwer S. (1975) The social significance of avian winter plumage variability. *Evolution* **29**, 593–610.
1.4.3

Rohwer S. (1977) Status signaling in Harris sparrows: some experiments in deception. *Behaviour* **61**, 107–129.
1.4.3, 5.4.1

Rohwer S. & Rohwer F.C. (1978) Status signalling in Harris sparrows: experimental deceptions achieved. *Animal Behaviour* **26**, 1012–1022.
1.4.3, 2.7, 5.4.1

Roth L.M. (1948) A study of mosquito behavior. An experimental laboratory study of the sexual behavior of *Aedes aegypti* (Linnaeus). *The American Midland Naturalist* **40**, 265–352.
4.3.1

Rowell T.E. & Hinde R.A. (1962) Vocal communication by the rhesus monkey (*Macaca mulatta*). *Proceedings of the Zoological Society of London* **138**, 279–294.
1.3.1

Russell C.J. & Bell C.C. (1978) Neuronal responses to electrosensory input in mormyrid *Valvula cerebelli*. *Journal of Neurophysiology* **41**, 1495–1510.
4.2.3

Rutowski R.L. (1978) The form and function of ascending flights in *Colias* butterflies. *Behavioral Ecology and Sociobiology* **3**, 163–172.
2.8.3

Ryan M.J., Tuttle M.D. & Taft L.K. (1981) The costs and benefits of frog chorusing behavior. *Behavioral Ecology and Sociobiology* **8**, 273–278.
3.6.2

Scheich H., Langner G. & Bonke D. (1979) Responsiveness of units in the auditory neostriatum of the guinea fowl to species specific calls and synthetic stimuli. II. Discrimination of iambus-like calls. *Journal of Comparative Physiology* **132**, 257–276.
4.5.2

Schleidt W.M. (1973) Tonic communication: continual effects of discrete signs in animal communication systems. *Journal of Theoretical Biology* **42**, 359–386.
1.3.1

Schmitt F.O. & Worden F.G. (ed.) (1974) *Neurosciences Third Study Program*. MIT Press, Cambridge, Mass.
4.5

Schneider D. (1965) Chemical sense communication in insects. *Symposia of the Society for Experimental Biology* **20**, 273–297.
4.2.1

Schneider D. (1969) Insect olfaction: deciphering system for chemical messages. *Science* **163**, 1031–1037.
4.2.1

Schneider D. (1974) The sex-attractant receptor of moths. *Scientific American* **231**, 28–35.
2.2.1

Schneider D., Block B.C., Boeckh J. & Priesner E. (1967) Die Reaktion der männlichen Seidenspinner auf Bombykol und seine Isomeren: Elektroantennogramm und Verhalten. *Zeitschrift für vergleichende Physiologie* **54**, 192–209
4.2.1

Schuijf A. (1975) Directional hearing of cod (*Gadus morhua*) under approximate free field conditions. *Journal of Comparative Physiology* **98**, 307–332.
4.6.2

Schuijf A. & Buduwala R.J.A. (1975) On the mechanism of directional hearing in cod (*Gadus morhua* L.). *Journal of Comparative Physiology* **98**, 333–343.
4.6.2

Schwartz J.J. & Wells K.D. (1983) An experimental study of acoustic interference beteen two species of neotropical treefrogs. *Animal Behaviour* **31**, 181–190.
3.5

Schwartzkopf J. (1949) Über Sitz und Leistung von Gehör und Vibrationssinn bei Vögeln. *Zeitschrift für vergleichende Physiologie* **31**, 527–608.
4.3.1

Searcy W.A. & Marler P. (1981) A test for responsiveness to song structure and programming in female sparrows. *Science* **213**, 926–928.
1.4.1

Sebeok T.A. (ed.) (1977) *How Animals Communicate*. Indiana University Press, Bloomington.
1.6, 2.13

Serpell J.A. (1981) Duets, greetings and triumph ceremonies: analogous displays in the parrot genus *Trichoglossus*. *Zeitschrift für Tierpsychologie* **55**, 268–283.
2.5

Seyfarth R.M. & Cheney D.L. (1980) The ontogeny of vervet monkey alarm calling behavior: a preliminary report. *Zeitschrift für Tierpsychologie* **54**, 37–56.
5.7

Seyfarth R.M., Cheney D.L. & Marler P. (1980) Vervet monkey alarm calls: semantic communication in a free-ranging primate. *Animal Behaviour* **28**, 1070–1094.
1.2.1, 2.9.1

Shalter M.D. (1978) Localisation of passerine seet and mobbing calls by goshawks and pygmy owls. *Zeitschrift für Tierpsychologie* **46**, 260–267.
3.4.3

Shannon C.E. & Weaver W. (1949) *The Mathematical Theory of Communication.* University of Illinois Press, Urbana.
2.11, 5.2.1

Shaw E.A.G. (1974) The external ear. In: *Handbook of Sensory Physiology* Vol. V/1 (ed. W.D. Keidel & W.D. Neff), pp. 455–490. Springer-Verlag, Berlin.
4.6.1

Shorey H.H. (1976) *Animal Communication by Pheromones.* Academic Press, New York.
3.2, 3.2.1, 3.2.3, 3.8

Shorey H.H. (1977) Pheromones. In: *How Animals Communicate* (ed. T.A. Sebeok), pp. 137–163. Indiana University Press, Bloomington.
1.4.2

Sibley C.G. (1957) The evolutionary taxonomic significance of sexual dimorphism and hybridization in birds. *Condor* **59**, 166–191.
5.2.2

Sigurjónsdóttir H. & Parker G.A. Dung fly struggles: evidence for assessment strategy. *Behavioral Ecology and Sociobiology* **8**, 219–230.
5.5.2, 5.5.4

Simmons J. (1973) The resolution of target range by echolocating bats. *Journal of the Acoustical Society of America* **54**, 157–173.
4.3.2

Simpson M.J.A. (1968) The display of the Siamese fighting fish, *Betta splendens. Animal Behaviour Monographs* **1**, 1–73.
1.4.3, 5.5.2

Slater P.J.B. (1973) Describing sequences of behavior. In: *Perspectives in Ethology* (ed. P.P.G. Bateson & P.H. Klopfer), pp. 131–153. Plenum Press, New York.
1.3.1, 1.6

Slater P.J.B. (1981) Chaffinch song repertoires: observations, experiments and a discussion of their significance. *Zeitschrift für Tierpsychologie* **56**, 1–24.
1.4.1

Smith D.G. (1972) The role of the epaulets in the red-winged blackbird (*Agelaius phoeniceus*) social system. *Behaviour* **41**, 251–268.
1.4.3

Smith R.J.F. (1976) Chemical communication as adaptation: alarm substance of fish. In: *Chemical Signals of Vertebrates* (ed. D. Müller-Schwarze & M.M. Mozzell), pp. 303–320. Plenum Press, New York.
1.4.2

Smith W.J. (1965) Message, meaning and context in ethology. *American Naturalist* **99**, 405–409.
1.2.1

Smith W.J. (1968) Message–meaning analyses. In: *Animal Communication* (ed. T.A. Sebeok), pp. 44–60. Indiana University Press, Bloomington.
1.2.1, 1.6, 5.2.1

Smith W.J. (1977) *The Behavior of Communicating.* Harvard University Press, Cambridge, Mass.
1.1, 1.2.1, 1.2, 2.13, 5.2.1

Sower L.L., Kaae R.S. & Shorey H.H. (1973) Sex pheromones of lepidoptera. XLI. Factors limiting potential distance of sex pheromone communication in *Trichoplusia ni. Annals of the Entomological Society of America* **66**, 1121–1122.
3.2.1

Stamps J.A. & Barlow G.W. (1973) Variation and stereotypy in the displays of *Anolis aeneus* (Sauria; Iguanidae). *Behaviour* **47**, 67–94.
   1.3.1, 5.2.2

Stevenson J.G., Hutchison R.E., Hutchison J.B., Bertram B.C.R. & Thorpe W.H. (1970) Individual recognition by auditory cues in the common tern (*Sterna hirundo*). *Nature* (London) **226**, 526–563.
   2.5

Stokes A.W. (1962) Agonistic behaviour among blue tits at a winter feeding station. *Behaviour* **19**, 118–138.
   1.3.1

Stout J.F. & Brass M.E. (1969) Aggressive communication by *Larus glaucescens*. Part II. Visual communication. *Behaviour* **34**, 42–54.
   1.4.3

Struhsaker T.T. (1967) Auditory communication among vervet monkeys (*Cercopithecus aethiops*). In: *Social Communication Among Primates* (ed. S.A. Altmann), pp. 281–324. University of Chicago Press, Chicago.
   1.2.1

Suga N. & O'Neill W.E. (1979) Auditory processing of echoes: representation of acoustic information from the environment in the bat cerebral cortex. In: *Animal Sonar Systems*. (ed. R.G. Busnel & J.F. Fish), pp. 589–611. NATO Advanced Studies Institute, Series A, Life Sciences Vol. 28. Plenum Press, New York.

Szabo T. (1974) Anatomy of specialized lateral line organs of electroreception. In: *Handbook of Sensory Physiology* Vol. III/3 (ed. A Fessard), pp. 13–58. Springer-Verlag, Berlin.
   4.2.3

Szabo T., Enger P.S. & Libouban S. (1979) Electrosensory systems in the mormyrid fish *Gnathonemus petersii*: special emphasis on the fast conducting pathway. *Journal de Physiologie* **75**, 409–420.
   4.2.3

Tansley K. (1965) *Vision in Vertebrates*. Chapman & Hall, London.
   3.3.4

Tavolga W.N., Popper A.N. & Fay R.R. (eds) (1981) *Hearing and Sound Communication in Fishes*. Springer-Verlag, Berlin.
   4.8

Thornhill R. (1979) Adaptive female-mimicking behavior in a scorpionfly. *Science* **205**, 412–414.
   2.8.3, 5.4.1

Thorpe W.H. (1972) Duetting and antiphonal song in birds. *Behaviour* Suppl. 18.
   2.5

Tinbergen N. (1952a) The curious behavior of the stickleback. *Scientific American* **187** (12), 22–26.
   5.2.2

Tinbergen N. (1952b) Derived activities; their causation, biological significance, origin and emancipation during evolution. *Quarterly Review of Biology* **27**, 1–32.
   2.8.4

Tinbergen N. (1953) *Social Behaviour in Animals*. Methuen, London.
   2.8.3

Tinbergen N. (1959) Comparative studies of the behaviour of gulls (Laridae): a progress report. *Behaviour* **15**, 1–70.
   1.2.2, 1.3.1

Tobin T.R. (1981) Pheromone orientation: role of internal control mechanisms. *Science* **214**, 1147–1149.
3.2.2

Tolman C.W. (1968) The role of the companion in social facilitation of animal behavior. In: *Social Facilitation and Imitative Behavior* (ed. E.C. Simmel, R.A. Hoppe & G.A. Milton), pp. 33–54. Allyn & Bacon, Boston.
1.1

Trivers R.L. (1972) Parental investment and sexual selection. In: *Sexual Selection and the Descent of Man* (ed. B. Campbell), pp. 136–179. Heinemann, London.
2.8.3

Tuttle M.D., Taft L.K. & Ryan M.J. (1982) Evasive behaviour of a frog in response to bat predation. *Animal Behaviour* **30**, 393–397.
3.6.2

Waldman B. (1981) Sibling recognition in toad tadpoles: the role of experience. *Zeitschrift für Tierpsychologie* **56**, 341–358.
2.6

Waldman B. & Adler K. (1979) Toad tadpoles associate preferentially with siblings. *Nature* (London) **282**, 611–613.
2.6

Walkowiak W. (1980) Sensitivity, range and temperature dependence of hearing in the grass frog and fire-bellied toad. *Behavioural Processes* **5**, 363–372.
3.6.1

Walls G.L. (1942) *The Vertebrate Eye and its Adaptive Radiation.* Hafner, New York.
3.3.5

Ward P. & Zahavi A. (1973) The importance of certain assemblages of birds as 'information centres' for food finding. *Ibis* **115**, 517–534.
2.9

Wasserman F.E. (1977) Interspecific acoustical interference in the white-throated sparrow *Zonotrichia albicollis*. *Animal Behaviour* **25**, 949–952.
1.4.1

Weeden J.S. & Falls J.B. (1959) Differential responses of male ovenbirds to recorded songs of neighboring and more distant individuals. *Auk* **76**, 343–351.
1.4.1

Wehner R. (1981) Spatial aspects of vision in arthropods. In: *Handbook of Sensory Physiology* Vol. VII/6c (ed. H. Autrum), pp. 287–616. Springer-Verlag, Berlin.
4.2.4

Wehrhaln C. (1979) Sex specific differences in the chasing behavior of houseflies (*Musca*). *Biological Cybernetics* **32**, 239–241.
4.2.4

Weidmann U. & Darley J.A. (1971) The role of the female in the social display of mallards. *Animal Behaviour* **19**, 287–298.
1.3.1

Wells K.D. (1980) Intraspecific and interspecific communication in the neotropical frog, *Hyla ebraccata*. *American Zoologist* **20**, 724.
3.5

West M.J., King A.P., Eastzer D.H. & Staddon J.E.R. (1979) A bioassay of isolate cowbird song. *Journal of Comparative and Physiological Psychology* **93**, 124–133.
1.4.1

White S.J. & White R.E.C. (1970) Individual voice production in gannets. *Behaviour* **37**, 40–54.
5.2.2

Wickler W. (1968) *Mimicry in Plants and Animals*. Weidenfeld & Nicholson, London.
2.10.2, 2.10.3

Wickler W. (1980) Vocal duetting and the pair bond. I. Coyness and partner commitment. A hypothesis. *Zeitschrift für Tierpsychologie* **52,** 201–209.
2.5

Wiepkema P.R. (1961) An ethological analysis of the reproductive behaviour of the bitterling (*Rhodeus amarus* Bloch). *Archives néerlandaises de Zoologie* **2,** 103–199.
1.3.1

Wiley R.H. (1973) The strut display of male sage grouse: a 'fixed' action pattern. *Behaviour* **47,** 129–152.
5.2.1

Wiley R.H. (1975) Multidimensional variation in an avian display: implications for social communication. *Science* **190,** 482–483.
5.2.2

Wiley R.H. (1976) Communication and spatial relationships in a colony of common grackles. *Animal Behaviour* **24,** 570–584.
5.2.2

Wiley R.H. & Richards D.G. (1978) Physical constraints on acoustic communication in the atmosphere: implications for the evolution of animal vocalizations. *Behavioral Ecology and Sociobiology* **3,** 69–94.
3.4.1, 3.4.2, 3.8

Wiley R.H. & Richards D.G. (1983) Adaptations for acoustic communication in birds: sound transmission and signal detection. In: *Ecology and Evolution of Acoustic Communication in Birds* (ed. D.E. Kroodsma & E.H. Miller). Academic Press, New York.
3.4.2, 5.2.2, 5.8

Wiley R.H. & Wiley M.S. (1977) Recognition of neighbors' duets by stripe-backed wrens *Compylorhynchus nuchalis*. *Behaviour* **62,** 10–34.
5.6.2

Wilson E.O. (1970) Chemical communication within animal species. In: *Chemical Ecology* (ed. E. Sondheimer & J.B. Simeone), pp. 133–155. Academic Press, New York.
3.2.1

Wilson E.O. (1971) *The Insect Societies*. Belknap Press, Cambridge, Mass.
2.4

Wilson E.O. & Bossert W.H. (1963) Chemical communication among animals. *Recent Progress in Hormone Research* **19,** 673–716.
3.2, 3.2.1, 3.2.2

Woodward P.M. (1964) *Probability and Information Theory with Applications to Radar*. Pergamon Press, Oxford.
4.3.2

Wrangham R.W. (1977) Feeding behaviour of chimpanzees in Gombe National Park, Tanzania. In: *Primate Ecology* (ed. T.H. Clutton-Brock), pp. 504–538. Academic Press, London.
1.1

Wu H.M.H., Holmes W.G., Medina S.R. & Sackett G.P. (1980) Kin preference in infant *Macaca nemestrina*. *Nature* (London) **285,** 225–227.
2.6

Wunderle J. (1978) Differential response of territorial yellowthroats to the songs of neighbors and near-neighbors. *Auk* **95**, 389–395.
1.4.1

Yasukawa K. (1981) Song repertoires in the red-winged blackbird (*Agelaius phoeniceus*): a test of the Beau Geste hypothesis. *Animal Behaviour* **29**, 114–125.
5.4.1

Young A.M. (1981) Temporal selection for communicatory optimization: the dawn–dusk chorus as an adaptation in tropical cicadas. *American Naturalist* **117**, 826–829.
3.4.2

Zahavi A. (1975) Mate selection – a selection for a handicap. *Journal of Theoretical Biology* **53**, 205–214.
5.4.2

Zipser B. & Bennett M.V.L. (1976) Interaction of electrosensory and electromotor signals in lateral line lobe of a mormyrid fish. *Journal of Neurophysiology* **39**, 713–721.
4.2.3

INDEX

217